Warmonger

Warmonger
Vladimir Putin's Imperial Wars

Alex J. Bellamy

© Alex J. Bellamy 2023

This book is copyright under the Berne Convention.
No reproduction without permission.
All rights reserved.

First published in 2023 by Agenda Publishing

Agenda Publishing Limited
PO Box 185
Newcastle upon Tyne
NE20 2DH
www.agendapub.com

ISBN 978-1-78821-646-3 (hardcover)
ISBN 978-1-78821-647-0 (paperback)

British Library Cataloguing-in-Publication Data
A catalogue record for this book is available from the British Library

Typeset by JS Typesetting Ltd, Porthcawl, Mid Glamorgan
Printed and bound in the UK by TJ Books Limited, Cornwall

Contents

Introduction		1
1	Collapse	11
2	Chechnya	37
3	Georgia	57
4	Ukraine I	81
5	Syria	107
6	Nagorno-Karabakh	129
7	Shadows	149
8	Ukraine II	169
Chronology		191
Index		193

imperialism 1 a policy of extending a country's power and influence through colonization, use of military force, or other means
Oxford Concise English Dictionary

Introduction

Russia's invasion of Ukraine in late February 2022 caught many people by surprise. It should not have. This was an imperial war long in the making. It was a war that fulfilled the long-held needs and aspirations of Russia's president, his allies and dependents, and a large section of Russian society that had become willing accomplices in and cheerleaders for the state's imperial project. To paraphrase the sociologist Charles Tilly: war made Vladimir Putin, and Vladimir Putin made war. This book shows how.

Leningrad

Vladimir Vladimirovich Putin was born in Leningrad, today's Saint Petersburg, in 1952. The city young Vladimir knew bore the scars of its gruelling siege by the Nazis during the Second World War, a rundown wreck of broken buildings and hungry people. Thousands of orphaned, disowned and impoverished children roamed streets ruled by violent gangs. Putin shares his slight stature with many others of his generation, for they are the children of those who suffered most. Indeed, his may owe something to his mother's malnourishment.

Young Vladimir was raised in a shabby and spartan tenement block where the living accommodation was small and essential services communal. The Putins were by no means the worst off. Theirs was a middling lifestyle for 1950s' Leningrad. Vladimir's father was a factory foreman. After a stint in the navy, he had served in an NKVD (secret police) unit operating behind enemy lines before transferring to the regular army where he was seriously injured. His paternal grandfather had reached an altogether more elevated status as Stalin's chef. He had served Lenin before that. Putin's mother worked in a factory too. Vladimir was the couple's third son, but his siblings were long dead before he was born. He was an unremarkable school student, but a resourceful and tough-minded player in a very rough city. He occasionally loitered with the street gangs but ultimately chose a different path for himself. Like young people on both sides of the Iron Curtain, Vladimir was captivated by spy movies. He wanted to serve the state, as his grandfather had. Judo club taught him the discipline and focus he needed. Legend has it that he presented himself to Leningrad's KGB headquarters and volunteered for service. The bemused officials turned him away and told him to apply again when he had obtained his law degree. That is exactly what he did. Law degree in hand, Putin joined the KGB in 1975.

On the surface, the 1970s were a good decade for the Soviets and a poor one for the Americans. Under Leonid Brezhnev's (1964–82) leadership, the Soviet Union lost its revolutionary zeal and gained a stolidness verging on the comatose. But it had forged order out of chaos. Buildings had been rebuilt. Factories reignited. Livelihoods improved. The Soviets led the space race and the missile race. The USSR put a space station into orbit. It had more nuclear weapons than anyone else. It was surrounded by a ring of communist states who owed allegiance to Moscow. Whenever those governments stepped out of line, Soviet might returned them to the fold, usually with ease. Budapest 1956. Prague 1968. Kabul 1979, however, was a different story. The worst excesses of Stalinist terror were long past. Dissidents were imprisoned now, but not shot. Mass deportations and purges lived on only in memories and suppressed archives. By Brezhnev's time, most ordinary Russians rarely felt the wrath of the authorities because of their political opinions. Why would they? Most Russians were proud of their government. The Soviet Union was an unchangeable fact of life. An insuperable powerhouse that had improved their lives and stamped itself on the world. In 1975 Vladimir Putin became part of that great machine of power. He could look forward to a long career in the KGB and a life of intrigue but also of pride, privilege and security.

But that isn't how things worked out.

INTRODUCTION

Dresden

Putin's first decade in the KGB proved less than glamour-filled. His time was mostly spent training and keeping tabs on foreigners in Leningrad. He had to wait a decade for his first foreign posting and the opportunity to use the German language skills he had excelled at since school. But when his long-anticipated foreign posting came in 1985, it wasn't to the Cold War's nerve centre of Berlin, but to the East German backwater of Dresden. One view is that as an unexceptional agent, he was assigned an unexceptionable posting. A posting so far removed from the world of international espionage that he frequently posed for public photos. An alternative view is that Putin handled Soviet aid to the Red Army Faction and other leftist terrorists operating in Western Europe. Both views might contain an element of truth.

Barely anyone realized that the mighty Soviet empire was on the brink of collapse. That the social, economic and technological advances everywhere visible rested on a combustible mix of oil revenues and debt. Collapse, when it came, was unexpected, swift and devastating – especially for Putin, a loyal servant of what he thought was the most powerful state on earth. East Germany was at the epicentre of the first wave of communist collapse. A few weeks after the Berlin Wall was breached in November 1989, a crowd of protestors descended on the headquarters of the hated East German secret police – the Stasi. A small group broke off from the main demonstration and headed for the KGB's office, situated in a large house nearby. They were met at the gate by an agitated Russian. Vladimir Putin warned that the officers inside the building were armed and would use their weapons if the protestors broke into the compound. Deciding that discretion was the better part of valour, the protestors withdrew and rejoined the main group at the Stasi building. An anxious Putin returned to his office and called a Russian tank unit stationed nearby for help. "We cannot do anything without orders from Moscow" came the frosty reply. "And Moscow is silent". Putin was stunned. How could Moscow, the greatest power on earth, be silent? And how could the Soviet army abandon them to their fate? Moscow did in fact break its silence soon afterwards and the army was despatched to protect the KGB's house. So it's not quite true that no one came to help the beleaguered Putin as the president later claimed. It's an apocryphal story though, in both senses of the word,

3

in that it is not entirely accurate, but nonetheless discloses a deeper truth. A truth about the speed of Soviet collapse and the bewilderment it caused for the self-identities, sense of worth, place and mission, security, livelihoods and worldviews of the millions of people whose lives were defined by a Soviet state they had assumed would last forever. Two weeks after the stand-off, Putin saw West German Chancellor Helmut Kohl visit Dresden. Received by a massive adoring crowd, Kohl praised Soviet leader Mikhail Gorbachev and hinted that East Germany might one day be reunited with the West.

Something else nobody expected was that the end of communism would not just tear the Soviet Union's East European satellites away from it but that it would destroy the Soviet Union itself. The Soviet Union lost an empire, Eastern Europe; But Russia lost an empire too, the Soviet Union itself. A little more than two years after an anxious KGB agent in Dresden was informed that "Moscow is silent", the Soviet Union was permanently silenced. Without ceremony, a few days before the end of 1991, the Soviet flag was hauled down from the Kremlin's roof, replaced by the Russian tricolour. But what sort of Russia would this new state be?

The Kremlin

It is 24 February 2022. The KGB agent from Dresden is president of the Russian Federation. Only 2 per cent of Russians had thought him presidential material when Boris Yeltsin had anointed him as his successor in August 1999. Yet just seven months later, Putin had won a crushing first-round election victory, winning almost double the number of votes of his nearest rival, the communist Gennady Zyuganov. His was a meteoric political rise with few precedents. Like the collapse of the USSR, unthinkable before it actually happened. It was war that had propelled Putin to the presidency. A bloody war of indiscriminate violence in Chechnya. Now President Putin was announcing a new war: the invasion of Ukraine. In a pre-recorded address to the nation, Putin honed-in on the problem that had long been at the forefront of his mind. Perhaps since his Dresden days. Everything "was clear and obvious" to him. It had all begun with the collapse of the Soviet Union. The collapse was a lesson, "that the paralysis of power and will is the first step towards complete degradation and oblivion". Back then, Russia had lost its bearings for just a moment, but that had been enough to throw the whole world into disarray. "Old treaties and agreements are no longer effective". NATO had expanded eastwards. Western governments sponsored separatist terrorism in Russia itself. Russia's neighbours, its "historical land", had become hostile towards Moscow. Ukraine especially, now a haven for "far-right nationalists and neo-Nazis" controlled by the US and intent on degrading and destroying Russia. Now he, Vladimir Putin, was going to put things right. "For our country, it is a matter of life and death, a matter of our historical future as a nation". There was sheer farce of course. Never has a leader declared war whilst also claiming that "we do not intend to impose anything on anyone by force". But there was chilling threat too. "I would now like to say something very important for those who may be tempted to interfere in these developments from the outside. No matter who tries to stand in our way or all the more so create threats for our country and our people, they must know that Russia will respond immediately, and the consequences will be such as you have never seen in your entire history".

As the president's address was being beamed to the nation, Russia's military was already invading Ukraine from the north, east and south. Putin, his advisors, and most pundits expected Ukraine's government to quickly collapse. But it didn't. After a month or so, the Ukrainians started retaking territory. When Russia lost the Battle of Kyiv and retreated, what the Ukrainians discovered in their wake was a massacre. In towns and villages like Bucha, Russian troops had gone house to house searching for men and older boys. Those they found were bound, hooded and shot. Russian soldiers also searched for women. Many women and girls were captured, tortured, raped and murdered. Their bodies often burned to conceal evidence of their suffering. More than 650 civilians were killed this way, their broken bodies discovered when the Ukrainian army liberated Bucha. This was war by atrocity.

Indiscriminate bombardment, the slaughter of civilians, widespread rape and sexual violence, the brutal torture of prisoners – all this was visited on Ukraine. Twenty-three years earlier the same horrors had been inflicted on Chechnya, by the same army directed by the same political leader. Atrocity is one of the defining features of the Russian way of warfare. It pre-dates Putin and is hard-wired into the Russian armed forces. National service in Russia is notoriously vicious – and deadly. Dozens of conscripts die at the hands of their comrades each year. Dozens more commit suicide. For all the glitz of its publicly paraded new weapons systems, the Russian armed forces still systematically brutalize their own to prepare them to kill, maim and torture their enemies without discrimination. The army's brutality mirrors Russian society's endemic problem with violence. Even after two decades of economic growth and political stability under Putin, Russia's murder rate is still higher than Afghanistan's, a third higher than Ukraine's, close to double that of the United States, and more than eight times that of the UK. For all its surface appearance of orderliness Russia's is an intensely violent society.

War

The invasion of Ukraine did not come out of the blue. It was the manifestation of a clearly articulated political project, albeit one developed piecemeal over several years. And it was merely the latest of a long line of situations where the Russian state had employed violence to realize that project. By the time Russian tanks began their roll towards Kyiv, Putin's Russia had already fought wars in Chechnya, Georgia, Syria and Ukraine. Its "peacekeepers" held territory in Moldova (Transnistria), Azerbaijan (Nagorno-Karabakh) and Georgia (Abkhazia and South Ossetia). Russian mercenaries fought for governments in Syria, South Sudan, Mali and the Central African Republic, and against the government in Libya. February 2022 wasn't even the first time Putin's army had invaded Ukraine. Russia had already occupied and annexed Crimea, and Russian forces and their allies held parts of the country's eastern Donbas region. It is hard to escape an obvious conclusion from all this military activity. That Putin's Russia wants its empire back and is willing to use war to get it. But Putin's vision of Russia's imperial destiny is not limited to what Russians call their "near abroad" – the ring of states once part of the Soviet Union that many Russians, including the president, think aren't really states at all. Control of an imperial space is just one part of a broader global struggle for status in which Russia seeks to offer itself as an alternative to American liberalism. This book argues that Russia has been at war with the West for years in pursuit of these twin goals but that it took the invasion of Ukraine in 2022 for the West to wake up to that fact.

Putin may be the conductor, but he does not act alone. He is powerful because he commands a vast state apparatus that shares his vision. Putin and the Russian state are codependents. The Russian state, its armed forces, security services, spooks and paramilitaries, its bureaucrats, politicians, political technologists, media operatives, television stars, filmmakers and functionaries are a willing orchestra. And we should not fool ourselves into thinking that ordinary Russians are secretly horrified by what the state does in their name. It is true that most people in Putin's Russia are fed an information diet consisting mostly, if not exclusively, of state propaganda. But even independent polling suggests most Russians support their president, his vision of Russia's identity and place in the world, and the warmongering that flows from that. In their book *Putin v. the People*, Samuel Greene and Graeme Robertson show how it is not just the state that is locked into codependency with Putin but the Russian people too. After the trauma of collapse, Putin sold the Russian people a vision of themselves that they liked. He told them that they were an exceptional people, triumphant against the Nazis, gifted in arts and culture, destined by the grace of God to incubate true moral values and be the "Third Rome", and that because of this they were ceaselessly attacked by outsiders – Nazis, Americans – and betrayed by those who owed everything to Mother Russia, the Ukrainians especially. Soviet collapse had traumatized them. The brief liberal experiment of the 1990s had failed them. Putin would restore them to greatness. Most Russians lapped it up. The rise of Putinism was thus largely consensual, based on a new social contract between people and president. Putin's popularity soared when Russia invaded Ukraine. Even the most independent of pollsters, the Levada Center, suggested 80 per cent of Russians supported the war, although they later concluded that the true figure was likely much lower. Still, most Russians, it seems, want their global status and their empire back. But it remains to be seen whether their imperial nostalgia can withstand the crushing blows being dealt to it by the courageous defence of Ukraine. After all, a regime built on war is simultaneously at its most vulnerable and most dangerous when war stops delivering and begins to shake a society's material and political foundations to the core.

This book argues that war was always central to Putin's project. Collapse, corruption and violence shaped the context of his rise from

Saint Petersburg to the heart of Boris Yeltsin's government. War in Chechnya propelled Putin to the presidency. War then proved central to his rule. It helped craft the new social contract between president and people, a contract grounded in a shared vision of Russian national identity and its place in the world. That vision is an inherently imperial one. The double-headed eagle that adorns the national coat of arms and the presidential standard is self-consciously imperial. First adopted by the Grand Dukes of Muscovy in the fifteenth century, the golden eagle is an explicit reference to Byzantium, and its adoption, a claim that Muscovy was the "Third Rome" (Byzantium being the second), the centre of a Christian empire. In 2016, Putin had an enormous statue of the tenth–eleventh-century Saint Vladimir erected immediately outside the Kremlin's walls, in Borovitskaya Square. The saint, Putin opined as he unveiled the statue, "has gone down forever in history as the unifier and defender of Russian lands". But Saint Vladimir – Volodymyr in his native Ukrainian – was a prince of Kyiv not Muscovy. Putin's vision of Russia itself reaches deep into the heart of Ukraine.

War fulfilled several major roles in sustaining this vision. Memory of the "Great Patriotic War" (Second World War) was made a glorious myth to unite the Russian people; militarism and military modernization become practical symbols of Russia's return to great power status; and the use of force became one of the principal tools for holding Russia's imperium together and ensuring that Moscow controlled the spaces beyond its borders it believed owed tribute to the imperial centre. Russia's 2008 invasion of Georgia was a moment rightly described by Ben Judah in his book *Fragile Empire* as "peak Putinism". There, Russia scored a seemingly crushing victory over a government that had wanted to forge a future for itself independent of Moscow. But resistance to the Russian imperium was not limited to Georgia. In 2014, Russia invaded Ukraine to punish it for pursuing a destiny separate to Moscow and simply annexed Crimea.

The invasion of Ukraine in 2022 needs to be understood as one part of a much longer imperial project fought to support three interrelated objectives: (1) to sustain Putin's incontestable grip on power; (2) to build or restore as much imperial control over Russia's neighbours as possible; and (3) by achieving the first two goals, re-establish Russia as a global superpower, a peer challenger to American hegemony capable of drawing others to it.

None of this arrived fully formed in the president's mind, or in that of an advisor or political technologist. It developed contingently and incrementally in response to the practical challenges of sustaining and legitimizing rule in twenty-first century Russia. Putin and his followers had their own ideas of course. Belief in the primacy of the state was a constant that reached back into Putin's early career. Russian nationalism too, though precisely what that meant crystallized over time to the point where Putin could set out Russia's imperial claim to Ukraine in a 7,000-word essay. From experience in Chechnya, Georgia, Crimea and Syria they learned that the West was weak and divided and that their enemies could be expected to fold in the face of Russian military power. The West, it is true, made a series of miscalculations. It allowed itself to be distracted by the "War on Terror" and took its eye off European security. It exhibited weakness. It evinced disunity in the form of Brexit, the rise of populism and the lunacy of Donald Trump. It believed its own rhetoric about the end of history and a new international politics. Self-proclaimed "realists" forgot basic lessons of geopolitics and counselled that the security of East European democracies be sacrificed and Putin appeased. Western governments hesitated to grant Ukraine, Georgia and Moldova the security guarantees they so desperately needed in the belief Russia could be satiated by concessions or socialized into a rules-based international order. Western governments continued to treat Russia as a normal power even as the Kremlin voraciously gobbled up chunks of territory from its neighbours, even as the Russian government came to think and act as if locked in a global struggle, a war in fact, with the West. Beyond the West, states like China, India, Brazil, South Africa and Indonesia that loudly virtue signal their warm embrace of state sovereignty and deep commitment to anti-colonialism said nothing and did nothing as Russia waged war to rebuild its empire. The UN membership even elected Russia – a serial perpetrator of war crimes and crimes against humanity – to its Human Rights Council. All this encouraged Putin to believe his own rhetoric. Hubris set in, hubris exposed by Ukrainian courage in 2022.

Through a tour of Russia's war zones, I shall show that war has always been central to Putin's project, hence the epithet "warmonger". According to the *Oxford English Dictionary*, a "warmonger" is "one who seeks to bring about or promote war". Putin has done both. The system of government he leads has done both. Warmongering is one of

Putinism's core instruments. It has been validated and supported by large sections of the Russian public. The book begins with some context, the collapse of the Soviet Union, and then starting with Putin's first war, in Chechnya, we move war by war until we get to Ukraine 2022.

This is a short book that tells an important story – the most important story for European security thus far this century. I want to tell it simply and concisely, so I have dispensed with the usual academic referencing. Instead, I refer to ideas borrowed from others directly in the text and have included a list of further reading at the end of each chapter for those who want greater depth or want to explore the evidence more directly. This book tries to raise our eyes, to see the bigger picture of what has been happening and why. I make no claim to special insight or inside knowledge. Debatable points are expressed as debateable points and the overall trajectory of the tale told here is not dependant on them. What this book tries to do is piece the jigsaw together, to show how everything hangs together so that the picture becomes clear. That picture tells the story of how rebuilding itself after the 1990s, Russia embraced an authoritarian politics and imperial view of itself and its place in the world that caused a series of bloody wars. Wars that have killed hundreds of thousands of people.

Further reading

Samuel A. Greene and Graeme B. Robertson, *Putin v. the People: The Perilous Politics of a Divided Russia*. Second edition. New Haven, CT: Yale University Press, 2022.

Nataliya Gevorkyan, Natalya Timakova and Andrei Kolesnikov, *First Person: An Astonishingly Frank Self-Portrait by Russia's President Vladimir Putin*. Translated by Catherine A. Fitzpatrick. New York: Public Affairs, 2000.

Masha Gessen, *The Man Without a Face: The Unlikely Rise of Vladimir Putin*. London: Granta, 2013.

Ben Judah, *Fragile Empire: How Russia Fell in and Out of Love With Vladimir Putin*. New Haven, CT: Yale University Press, 2013.

Serhii Plokhy, *Lost Kingdom: A History of Russian Nationalism from Ivan the Great to Vladimir Putin*. London: Penguin, 2017.

1
Collapse

To start, we need to understand why and how the Soviet Union collapsed and how the trauma of collapse shaped how Putin, his allies and millions of Russians came to view the world. Putin described Soviet collapse as "the greatest geopolitical catastrophe of the century". "Tens of millions of our fellow citizens and countrymen found themselves beyond the fringes of Russian territory". An "epidemic of collapse" as he called it, spilled into Russia, threatening its very existence. The 1990s are remembered as a traumatic time for most Russians in which quality of life and rule of law fell apart. Savings and job security evaporated. Rates of poverty, alcoholism and mortality increased. Average life expectancy fell by nearly five years. Russia's population declined. Putin came to power as prime minister at the end of that turbulent decade in 1999 and became president the following year. His presidency was defined by that backdrop; Putinism was all about reversing that catastrophe and restoring Russian pride by rebuilding the state to its former glory. What that meant exactly evolved over time, but Putinism always held Russia of the 1990s as its antithesis. Humiliation, decay, poverty and death are indelibly connected to state weakness, liberalism, democracy, and all things "Western" in the Putinist view of politics.

The collapse of communism sparked violent conflict and Russia used war, and proxies, to protect its interests. The collapse of the USSR unleashed powerful centrifugal forces that Moscow struggled to contain. Even then, Soviet and then Russian governments exhibited an interest in holding onto as much of the USSR as possible, especially those areas where Russians, Russian-speakers, and Russian allies lived, under Moscow's influence. They used force where they could, force moderated principally not by morality or political intent but by crippling incapacity. A will to fight unmatched by the capacity to do so was a recurrent theme, which stretched from Lithuania to the failed August 1991 coup, to Chechnya. Only against the unarmed, the very lightly armed, or other elements of its own enormous military institution did the Soviet and then Russian armies enjoy much success in the 1990s. But it was not for want of trying.

The West responded hesitantly, made anxious by uncertainty and risk. Western governments seemed reluctant to invest hope and cash in the new states that emerged from the rubble of Soviet empire. They banked on Yeltsin's reforms instead. But however courageous a political fighter, however noble his apparent commitment to democracy, Yeltsin had no more idea of how to reform Russia than Gorbachev. His economic "shock therapy" failed badly. To get a sense of just how badly, by the end of the 1990s, most of Eastern Europe's post-communist states were rebounding. By 1995 Poland's GDP per capita was nearly double what it had been in 1988. Hungary's rebound was slower but by 1995 it had added around 20 per cent to its 1988 GDP per capita. Romania's escape from the erratic despotism of Nicolae Ceaușescu was more troubled still. By 1995, it had not yet recovered to 1988 levels of wealth, less than half of Russia's. Its economy rebounded only at the end of the 1990s. Yet even poor Romania fared better than Russia in the 1990s. By 1995, Russia's GDP per capita was a third lower than it had been in 1988. By the time Putin became prime minister in 1999, Russia's GDP per capita was lower than Romania's.

Chaotic reform and disastrous performance encouraged opposition to which Yeltsin responded with authoritarian zeal. He employed tanks against parliament in 1993, attacked Chechnya in 1994, and permitted the wholesale looting of national wealth in return for election victory in 1996. Yet because his principal opponents were communists and far-right nationalists, Yeltsin remained the West's preferred president. Few retained much faith, however, in Yeltsin's ability or his commitment to democracy. This loss of faith goes some way to explaining why no one seriously contemplated a "Marshall Plan"-style economic bailout for post-Soviet Russia. Yeltsin's torrid legacy was an unhealthy and dispirited people; a broken and corrupted state; and a web of unresolved armed conflicts. It was from this that Putinism emerged, offering Russians a new social contract: a promise of improved lives and restored dignity in return for their unbending allegiance.

Gorbachev

The Politburo understood the Soviet Union was in trouble long before it elected Mikhail Gorbachev to be general secretary of the Communist Party in March 1985. Decades of agricultural collectivization had left the country unable to reliably feed itself. Decades of industrial central planning had left it unable to supply Soviet consumers with the goods they wanted. Both sectors exhibited only a fraction of the efficiency achieved in the West. Up to a third of all Soviet industries made a loss. Yet this failing economy had to support not just the Soviet Union, but world communism too. The legitimacy of communism in Eastern Europe was fortified by vast transfers of wealth from the USSR, provided mainly in the form of cut-price oil and gas. The East Europeans were supposed to supply manufactured goods in return, but it was a bargain they rarely kept. As a result, most Warsaw Pact countries enjoyed a standard of living considerably better than the Soviet Union, although that could never wholly compensate for the fact Poles, Hungarians, Czechs, Slovaks and East Germans were ruled by governments imposed on them by Moscow. Nor could it reduce the obvious gaps between their standard of living and that enjoyed on the other side of the Iron Curtain in Western Europe.

Propping up world communism was a costly business. From Cuba to Vietnam, communists looked to the Soviet Union for help. Soviet money, credits, industrial goods and arms flowed out, the price of waging a global ideological struggle against American-backed capitalism. The Soviet arms industry had to not just keep up with the West but outpace it. The USSR needed more tanks, more ships, and most importantly and expensively of all, more nuclear weapons. All that was hugely expensive.

A tipping point came in 1979 when the USSR invaded Afghanistan to protect Mohammed Najibullah's "people's democracy" from Islamist Mujahideen. Confronting an enemy they couldn't see and didn't understand, Soviet forces resorted to using force indiscriminately. They routinely killed and raped civilians to punish or deter support for the Mujahideen. They targeted children, leaving booby trapped toys for them to pick up and detonate. Sometimes, Soviet armoured vehicles surrounded villages and pummelled them to rubble. Conservative estimates suggest more than 400,000 Afghan civilians died at Soviet hands.

Many put that figure closer to one million. Yet none of that brought the Afghans to heal. All the while, the Mujahideen inflicted a terrible toll on the Soviet army. War in Afghanistan cost the lives of 14,000 Soviet soldiers and drained the economy still further.

The Soviet Union, however, had one great asset: the world's largest proven deposits of oil and gas. Eighty per cent of Soviet foreign currency earnings came from the sale of natural resources which meant that the USSR could just about balance the books while global energy prices were high. The 1970s oil shocks were a particular boon. In June 1973, oil cost $23 per barrel, by June 1980, it was $137. Rents earned from natural resources funded exorbitant military spending, subsidized communist governments, and covered obvious gaps in the ailing economy by, for instance, paying for imported grain. It was an inefficient way of using the income, but it kept the system going. The stability and power projected by Brezhnev's regime rested on the unstable foundation of global energy prices not on the ideological wisdom of communism.

Yuri Andropov, the former head of the KGB who succeeded Brezhnev on his death in 1982, understood this well. Oil prices were falling (down to $104 per barrel in October 1982) exposing cracks in the system. The US meanwhile had replaced the placid Jimmy Carter with the anything-but-placid Ronald Reagan. Reagan turned up the heat on the ideological battle, branding the Soviet Union an "evil empire" in 1983. He also ramped up US defence spending, promising to match the Soviet Union in the nuclear race and develop a missile defence system, the so-called "Star Wars" initiative, to neuter the Soviet arsenal. The US sent billions of dollars' worth of arms and other aid to those battling communism worldwide, most especially to the Mujahideen in Afghanistan. The cost of sustaining world communism thus increased just as the Soviet Union's ability to pay for it declined.

Andropov understood the structural weaknesses and counselled gradual reform. But he was ill when he assumed power and died 18 months later. The old guard tried to stem reformism by electing the reliably Brezhnevian Konstantin Chernenko as leader. At 73 years old, Chernenko had never had an original idea. He lasted only 13 months before dying in office. Yet even Chernenko understood that the system needed reform and so as his demise approached, he engineered his former rival, a young reformist called Mikhail Gorbachev, into an unassailable political position so that when he died, in March 1985,

Gorbachev replaced him. The world oil price was now $74 a barrel. Twelve months later it was just $27. The Soviet Union was running out of money. To plug the gaps, East European governments borrowed billions from Western banks promising industrial reform would help them pay it back. By the end of the decade, Gorbachev himself was begging Western governments for loans.

It was against this backdrop that Gorbachev launched his great reform programme, *perestroika* ("reconstruction"). The new leader was a Leninist not a liberal at heart. He wanted to save Marxism-Leninism by building a more sustainable "humane socialism", not destroy it. Like Khrushchev before him, Gorbachev believed Stalin had deviated from Lenin's path and that he could revive that original Bolshevik vision. He thought the system's legitimacy could be fortified by managed democratization. Gorbachev's politics became more liberal with time, but only because piecemeal reform was worsening the USSR's difficulties. Before his political end, Gorbachev began backsliding to such an extent that his political ally, foreign minister Eduard Shevardnadze, resigned.

Perestroika advanced three principal initiatives to repair the economy: firms were permitted to sell surplus produce on the open market; state ownership would be gradually replaced by the collective ownership envisioned by Lenin; and the state would lose its monopoly on international trade and foreign investment would be encouraged. It didn't work. Cooperatives and private firms sprung up alongside state enterprises to sell off surplus produce, but they didn't produce anything themselves. Instead, communist bureaucrats (*nomenklatura*) set themselves up as privateers and pocketed the profits. Since the government neither relaxed pricing restrictions nor took steps to reform the country's massive and decrepit industrial sector the economy's structural weaknesses remained.

To cut costs, it was imperative the Kremlin reduce its military spending and stop subsidizing world communism. Gorbachev withdrew Soviet forces from Afghanistan. He moved to end the Cold War by negotiating with Reagan a series of treaties to reduce nuclear and conventional stockpiles and end the arms race. And he renegotiated the Soviet Union's relationship with Eastern Europe. The "Brezhnev doctrine" was scrapped. No longer would the Soviet Union use its military might to prop up ailing communist governments elsewhere as it had done in Hungary (1956) and Czechoslovakia (1968). Economic subsidies and cheap energy were wound back. In 1988 Gorbachev announced the unilateral withdrawal of half a million Soviet troops from Eastern Europe. The USSR couldn't afford to keep them there, and he hoped this gesture might encourage the US to accept broader arms limitations that would help the Soviet budget.

Perestroika had a sibling: *glasnost*. An afterthought not a twin, it was catapulted to life by the economy's sluggish response to reform and the 1986 nuclear disaster in Chernobyl, northern Ukraine. To Gorbachev, Chernobyl was a microcosm of all that ailed the Soviet system. Decrepit infrastructure and incompetent managers caused a nuclear meltdown, the consequences of which the authorities tried to hide from the public. To survive, Gorbachev decided, the system must become accountable. Initially intended as a vehicle for bureaucratic transparency, *glasnost* became a byword for freedom of speech. It opened the door to the re-examination of Soviet history and its dark Stalinist past; allowed national groups to speak more openly about past genocides and abuses and present inequalities and discrimination. *Glasnost* brought into light the abuses, iniquities, contradictions, corruptions and failings of the Soviet system. It didn't assist *perestroika*, it killed it.

Since they didn't tackle the problems at source, Gorbachev's economic reforms didn't work. Production declined whilst *Glasnost* sent expectations soaring. The gap between what industry produced and what consumers wanted widened to a chasm. With the oil price catastrophically low, the Soviet Union was forced to export what manufactured goods it did produce to earn the foreign currency needed to import food. Shelves were empty, basic commodities difficult to source. Food rationing was introduced. People queued for hours to get bread. A tax on vodka, a desperate bid to raise cash and reduce alcoholism, simply alienated people further and drove the staple into the informal economy. Meanwhile, *Glasnost* allowed an avalanche of grievances to pour out. Opposition to Gorbachev's reforms grew. Hardliners wanted to turn the clock back; radicals wanted to run faster; and nationalists wanted to redraw boundaries or get out of the Soviet Union entirely.

As the pressure mounted, Gorbachev tried to circumvent the Party by appealing directly to the public for support. He believed most Soviet citizens still supported the state, humane socialism, and his reforms and thus tried to shift political authority away from the Communist Party towards new democratic institutions to inject legitimacy into the state.

He would sit atop those institutions as president of the Soviet Union not general secretary of the Communist Party. Gorbachev hoped that managed democratization would forge a broad coalition of support for reforming the Soviet Union. It didn't.

The centrepiece of political reform were the first free elections in the USSR's history. They were not free in the Western sense since all the candidates were members of the Communist Party, but the 1989 election to the Congress of People's Deputies allowed voters to choose between candidates holding different positions. The result did not go the way the government expected. Moscow, Leningrad and Kyiv rejected the party's preferred candidates. In Moscow, for example, an outspoken reformer called Boris Yeltsin was elected with 90 per cent of the vote. Indeed, right across the Soviet Union, people elected deputies who privileged their republic ahead of the Union. The nationalism of non-Russian peoples, permitted by *Glasnost*, now found a voice inside the corridors of Soviet power alongside the Russian nationalism there since Stalin. A caucus of 300 reformist deputies, nominally led by Yeltsin, pressed for speedier and more radical reform.

Yeltsin's election was especially fateful since he and Gorbachev had a history. Gorbachev had elevated Yeltsin to the position of secretary of the Moscow Communist Party, a non-voting position inside the Politburo. Yet, frustrated at the slow pace of reform, in 1987 Yeltsin had done something no Politburo member had ever done before and resigned. Gorbachev retaliated by denouncing Yeltsin to the Moscow Communists. Devastated, Yeltsin attempted suicide by plunging scissors into his chest. But Gorbachev wasn't finished. The leader summoned Yeltsin from his hospital bed for a ritual humiliation before the Party leadership, after which he was fired. Yeltsin never forgave Gorbachev's cruelty. Exacting revenge loomed large in his mind.

The *coup de grâce*, however, was delivered by Gorbachev himself. In March 1990, Gorbachev pushed through an amendment to Article 6 of the Soviet Constitution, abolishing the primacy of the Communist Party and thereby permitting multiparty elections. The move was intended to reinforce the transfer of political authority from the Party to the Soviet Union's reformed political institutions. What it did, however, was decapitate the state. Since the Bolshevik Revolution, the Soviet government had ruled through the Communist Party. Moscow controlled the nominally autonomous Soviet republics through the Party. From junior bureaucrats to republic leaders, the Party centre dictated who governed and on what terms. The Party determined who got what job, who got promoted, and who got sidelined. It dictated what every public servant could think, say and do. In a land without private property, everyone's living depended on the Party. Apartments, country dachas, coastal villas and cars were the gift of the Party. Rewards for right thought and faithful service could be granted and taken away. These were the mechanisms through which the Soviet centre controlled the periphery, through the Party not the state. Now that was gone and the republics could decide things for themselves.

They lost no time in doing so. Elections for republican and local parliaments (soviets) held between February and October 1990 brought popular national movements to power in Estonia, Latvia, Lithuania, Armenia, Georgia and Russia itself. These new democratic parliaments owed their allegiance to their republics and nations, not to the Soviet Union. Lithuania, Latvia, Estonia and Armenia quickly declared independence. Instead of legitimizing the Soviet state, democratization had empowered those determined to challenge its very existence.

The unravelling

East European economies stalled and indebtedness spiralled in the 1980s. Gorbachev's reforms compounded their problems. Economic subsidies, cheap resources, and the "Brezhnev doctrine's" promise that the Soviet military would bail them out of trouble, were all withdrawn. Gorbachev's message was that from now on they would have to sort out their own problems. It took a while for the message to get out, but when it did revolution quickly followed.

Most Poles had remained unreconciled to their communist government. They saw it as a foreign imposition first established by Stalin's pact with Hitler and still now enforced by the barrel of the gun. Unrest was common. Polish governments held onto power by shuttling between repression and concession. In 1980, a wave of strikes led by the trade union Solidarity occasioned both. The union's leadership was imprisoned while the government borrowed heavily from the West to subsidize better access to consumer goods. Yet the agitation and the strikes continued.

Things were no better in East Germany. Sovietization had never sat comfortably there either. In 1953, discontent provoked a wave of strikes and protests. Soviet tanks were called in to suppress it and more than one hundred people were killed, a foretaste of what was to come in Budapest (1956) and Prague (1968). With order restored, East Germany settled into a new pattern of government that combined a hardline security state with an economic policy more liberal than the USSR. That bargain, an uneasy marriage of consumer socialism and doctrinal conservatism, persisted into the 1980s under the watchful eye (from 1971) of Erich Honecker. The bargain limited outright protest but did not redress the regime's underlying legitimacy problem, a problem exacerbated by the flourishing of its capitalist kin, West Germany.

Hungary went the same way. Granted more economic freedoms than other Warsaw Pact countries after the violent suppression of the 1956 uprising, Hungary's standard of living was relatively high but the demand for consumer goods higher still. The government took on debt to fulfil those demands, becoming the most heavily indebted of all by 1987. Romania was heavily indebted too, but its dictator Nicolae Ceaușescu resolved to repaying the debt by squeezing the standard of living. As a result, Romania's GDP per capita was 25 per cent lower in 1988 than it had been in 1982. Like East Germany, the state in Czechoslovakia governed a population that deeply resented it, a resentment deepened by the USSR's 1968 military intervention to repress Prague's reformist government. The government Moscow imposed aimed for as little change as possible and banked on Soviet military and economic aid. It also borrowed heavily in a vain effort to buy popular legitimacy.

The revolutions began in Polamd. In 1989, a new wave of strikes led by Solidarity brought industry to a standstill at precisely the time the government needed to sell goods to service its mounting debts. Soviet-assisted suppression no longer an option, the government offered limited concessions, including multiparty elections, expecting it could maintain a toehold on power and gain some legitimacy. Instead, Solidarity won a crushing victory, winning 92 of the 100 contested seats in the Senate and 160 of 161 contested seats in the lower house. By the end of August, Poland's government was led by dissident intellectual and Solidarity member, Tadeusz Mazowiecki. The following year, Solidarity's leader Lech Wałęsa was elected president. Communism evaporated in the space of a few weeks.

Hungary was next. Emboldened by the elections in Poland, Hungarians took to the streets demanding their own. Arguably the most liberal of the Warsaw Pact's governments, the Hungarian regime relented. The result was the same as in Poland. By the end of October, Hungary was no longer ruled by communists.

Honecker's East Germany intended to put up a fight. There, the leadership instructed the notorious secret police, the Stasi, to prevent an uprising. People feared the police would open fire on them if they protested, but they turned out anyway, demanding their own elections. Half a million protested in central Berlin on 4 November 1989. Tens of thousands more East Germans took advantage of liberalization in Hungary to flee to the West. Still, Honecker contemplated using force to crack down, but he was caught in a bind. Gorbachev made it clear the Soviet army would not help; it was equally clear that the economy needed bailing out by West Germany. Using force on protestors would make that impossible. The crucial moment came on 9 November when crowds gathered at the Berlin Wall. Unsure of their orders and uncertain of their cause, the East German border guards stood aside. The wall was torn down. Less than a year later, East Germany itself was no more, replaced by a reunified Germany.

Czechoslovakia went the same way a couple of weeks later. Protests. Elections. End of communist rule. Ceaușescu put up a fight in Romania. Protests turned to riots. The feared Securitate opened fire, killing more than 600 but without the support of the army or the Soviet Union, the dictator's days were numbered. He was captured with his wife and shot.

The Warsaw Pact endured another 15 months. Gorbachev gamely hoped to refashion it along the lines of other security alliances, but the new democratic governments of Eastern Europe were eager to untether themselves from the Soviet Union. The Eastern bloc was formally dissolved in February 1991.

The pace of communism's collapse in Eastern Europe shocked the Kremlin. As did the uniform rejection of "humane socialism". The revolutions shattered the fiction that what people wanted was a better form of socialism, not a turn to capitalism. Still, perhaps the tumult could be explained by nationalism, perhaps as a reaction against Soviet imposition not socialist doctrine. That is what Gorbachev and his diminishing coterie of allies told themselves at the turn of the 1990s. But as they searched for an explanation, the same combination of economic collapse and political opening that had unravelled the Warsaw Pact was already unravelling the USSR itself. For all the problems, few saw the unravelling coming. The USSR was still a major power. It might struggle to put bread on the table, but it kept a space station in orbit. Living conditions, though poor by Western standards, were still better than in most parts of the world. Unemployment was low. Protests typically small and infrequent. Yet when it came, the unravelling was quick, bewildering and – ultimately – all-consuming.

To understand how the Soviet Union's unravelling caused a string of civil wars, why they occurred where they did, and how this set the terms for Russian imperial revival under Putin, we need to first understand Soviet federalism. Think of Soviet federalism as being like a Russian doll: a layered system of government designed to foster competition between and within its constituent parts so that each was kept in thrall to the centre. The building blocks were the Soviet Socialist Republics (SSRs). Theoretically sovereign and self-governing, each SSR was connected to a titular nation (e.g. Russian, Ukrainian, Latvian). Few, if any, were ethnically homogenous, however. Inside some SSRs were two other layers. Autonomous Soviet Socialist Republics (ASSRs) were self-governing entities within SSRs. These too were tied to a titular nation, one that was different to that of the SSR it was nestled within. For example, Abkhazia was an ASSR in Georgia tied to the Abkhaz nation; Chechnya in Russia was part of the Checheno-Ingush ASSR tied to the Chechen and Ingush peoples. Below ASSRs were Autonomous Oblasts (AOs). Also self-governing provinces within SSRs, these were not necessarily tied by name to a titular nation but often comprised national majorities different to that of the titular SSR. Nagorno-Karabakh, for example, was an Armenian majority AO within Azerbaijan SSR.

It was in Nagorno-Karabakh that the first signs of violent unravelling appeared. Three quarters of the population there was Armenian and in some places the distance between the oblast and Armenia itself is only five kilometres. For a brief period after the 1917 revolution, the area was administered by Armenia but was transferred by Stalin to Azerbaijan for reasons that are still debated. Immediately before that, Armenians had been subjected to a genocide by the Ottoman Turks that had claimed the lives of up to 1.5 million people. The separation of Nagorno-Karabakh from Armenia was a persistent source of discontent among Armenians. So it was not at all surprising that it rose to prominence with the arrival of *Glasnost*. Azerbaijanis in Nagorno-Karabakh were harassed, coerced, and forcibly displaced by Armenian activists. In February 1988, the AO's Armenian authorities demanded the oblast be transferred to Armenia SSR. The Politburo demurred. One million Armenians protested on the streets of Yerevan. Moscow sent the army to restore order. A couple of days later, an Azerbaijani mob, some (but by no means all) of them people forced from their homes in Nagorno-Karabakh, unleashed an anti-Armenian pogrom in Sumgait, a town about 30 km north of Baku. Armenians were attacked with knives, clubs and rocks. Dozens were raped. Properties burned. Official records show that 32 Armenians were killed, but Armenian sources put that figure at over 200. Understandably, the authorities in Yerevan saw immediate parallels with the 1915 genocide and demanded an urgent response from Moscow and Baku. They quickly concluded that the Soviet authorities were failing to protect Armenians living in Azerbaijan and resolved they would have to do it themselves. That would mean decoupling from the USSR and intervening in Azerbaijan.

The Armenian Supreme Soviet set its stall against both Moscow and Baku on 15 June 1988 by demanding the transfer of Nagorno-Karabakh. Through the summer, Azerbaijanis and Armenians conducted tit-for-tat killings and raids which escalated into low-level war between armed communities. The effect was a forced population transfer – Azerbaijanis out of Nagorno-Karabakh, Armenians out of Azerbaijan. The Soviet army tried to maintain order but proved incapable. One clash between soldiers and Karabakh Armenians left more than 60 injured and tore the last sinews of Yerevan's fidelity to the Soviet Union. By the following summer, Armenians in Nagorno-Karabakh had established an armed militia, backed by Yerevan. Moscow tried to restore order by assuming direct control of the oblast but without success. Armenians simply established their own parallel institutions. Meanwhile, angered that Moscow had taken control away from Baku, Azerbaijanis protested forcing the Kremlin to backtrack. In November 1989, it returned nominal control to Baku but since the Azerbaijan SSR had no authority and little presence inside Nagorno-Karabakh what was left there was essentially a 6,000 strong Soviet army of occupation. In January 1990, Azerbaijanis driven out of Nagorno-Karabakh inspired another anti-Armenian pogrom, this time in Baku. Mobs went street to street, killing, beating and raping. Armenians fled, many to the harbour where they were evacuated by ships arranged by the Soviet army. By the time the army stepped in to restore order, more than 100 people were dead, 1,000 injured, and almost the entire Armenian population of Baku, at least 13,000 people, were gone.

Soon afterwards, Soviet and Azerbaijani forces moved to restore order in Nagorno-Karabakh. There they found an entrenched partisan army willing to resist. They conducted "cleansing" operations to weed out Armenian insurgents and displace the villagers that supported them. But although the offensive certainly dented the Armenian insurgency, these brutal tactics only hardened Armenian resolve. Yet, it may have succeeded had the "Koltso" (Ring) operation not collapsed as a result of events in Moscow in August 1991.

Georgia was also vulnerable to the effects of Soviet unravelling. Located in the south Caucuses, Georgia housed an autonomous republic, Abkhazia, and two autonomous oblasts, South Ossetia and Adjara. Abkhaz comprised less than 20 per cent of Abkhazia's population (45 per cent of which was Georgian) but the region, which abuts the Black Sea, held a special place in the hearts of the communist elite as one of their favourite holiday destinations. South Ossetia also presented a problem since the majority Ossetians shared their nationality and a border with the Ossetians of North Ossetia, an autonomous oblast in Russia SSR. Like elsewhere, *Glasnost* stimulated Georgian nationalism and in November 1988 around 200,000 protestors took to the streets of Tbilisi demanding that republican law take precedence over Soviet law. Georgian leaders meanwhile complained about the special privileges enjoyed by the Abkhaz. Inevitably, the Abkhaz responded with their own protests demanding secession and incorporation into the Soviet Union. That, in turn, provoked an escalation of Georgian agitation.

In early April 1989, tens of thousands of Georgians protested on the streets of Tbilisi. Unable to control the situation, the authorities asked the Soviet army to intervene. Soldiers and armoured personnel carriers moved in, using noxious gas to disperse the demonstrators. Amidst the chaos, at least 19 protestors were killed, including a 16-year-old girl beaten to death by soldiers. More than 100 were injured. The violence provoked a shockwave rupturing relations between Moscow and Tbilisi. Moscow blamed the protestors and deposed the Georgian leadership but when a Congress of Deputies investigation found Soviet authorities responsible, Gorbachev changed his mind and blamed the army. The episode fed his doubts about the utility of using force to quell protests. The impact on Georgia was even more pronounced. Struggling to build legitimacy, the newly installed government bowed to popular demands and freed Georgian nationalists from jail and mandated the teaching of Georgian language in schools. It also legislated the precedence of Georgian law over Soviet law. Abkhazia reacted by proclaiming itself a sovereign republic and asking to join the Soviet Union. South Ossetia declared its wish to unite with North Ossetia within Russia. The seeds of future wars were sown. Georgia's first multiparty election, held in October 1990, brought a coalition of liberals and nationalists to power, headed by former dissident Zviad Gamsakhurdia, hardening the dividing lines between Georgia's centre and its autonomous regions.

Meanwhile, Gorbachev remained determined to hold as much of the Union together as he could. In doing so, his government poured fuel on the flames and stored up trouble for the future. In April 1990 his government legislated that in the event of an SSR electing to secede from the Soviet Union, its ASSRs and AOs would have the right to choose by referendum to remain a part of the Union. That is, ASSRs and AOs like Abkhazia and South Ossetia could secede from the SSR and join the Soviet Union. The intent was to deter secession, but the effect was to throw open the question of the SSRs' internal borders and increase the number and type of political entities that felt entitled to pursue independence. The idea that Russian and pro-Russian peoples could switch their allegiance from their own state to Russia, and by doing so earn themselves Moscow's help, became central to Putin's imperial vision of *Russkiy Mir* – a "Russian world" extending beyond the formal borders of the Russian Federation. In September 1990, South Ossetia appealed to remain in the Soviet Union and in December held a referendum. Gamsakhurdia's government responded by blockading the region, cutting off its electricity and gas, and – in January 1991 – deploying 5,000 Georgian militia and members of a newly formed national guard to impose Tbilisi's authority. This ragtag militia proved capable only of terrorizing civilians and looting property. When an attempt to backtrack by offering compromises failed, the national guard tried again in September but with the same effect.

In November 1990, Gorbachev attempted to stave off the slide towards secessionism by establishing new terms for the Union. He proposed a new union treaty that would decentralize power to the republics. The proposal won the assent of 70 per cent of the electorate in a March 1991 referendum, convincing Gorbachev that the majority was still with him. However, the referendum wasn't conducted in either the Baltic states or Georgia, where republican governments had already declared their independence and were refusing to recognize the supremacy of Soviet law. Voters elsewhere may have preferred the new model to the old, but that didn't mean they were committed to either. Disillusionment with Soviet reform grew with the food queues. Things came to a head in the Lithuanian capital, Vilnius, in January 1991. Gorbachev ordered the Lithuanian SSR to restore the Soviet Constitution and when it refused, Soviet troops moved in to take control of key institutions. Tanks and troops were met by thousands of protestors. Ordered to take the television tower, the soldiers opened fire and killed 14. Ordered to take the parliament building, they found their path blocked by up to 50,000 protestors. A moment of decision had arrived. Not for the last time, it was Yeltsin – never one to to miss an opportunity to contrast himself with Gorbachev – who acted decisively, forcing a moment of confrontation between Russia and the Soviet Union as he denounced the killings at the television tower and proclaimed Russia's support for the Lithuanians. Uncomfortable with the use of force, opposed by republican leaders like Yeltsin, and in a position where he could ill afford to alienate the West, Gorbachev buckled and the crackdown petered out.

The reality was that Gorbachev was in a vice-like grip from which he could not break free. Popular sentiment was almost everywhere slipping away from the Soviet centre towards republican leaderships. On the other side, conservatives were convinced things had gone too far and that Gorbachev's reckless policies were endangering the Union itself. The Soviet Union had entered its endgame.

On 19 August 1991, the day before the new union treaty was to be signed into law, conservatives led by Vice-President Gennady Yanaev, Prime Minister Valentin Pavlov, and KGB Chairman Vladimir Kriuchkov, sprung a coup. Gorbachev, who was holidaying at his dacha in Crimea was confined to quarters and cut off. The conspirators demanded his resignation as tanks rolled onto the streets of Moscow. But the putsch was badly planned, almost comically so. Perhaps its leaders had expected everyone to fall into line; perhaps they were simply inept. For whatever reason, the plotters failed to even try to deal with their republican and radical foes. Most notably, no attempt was made to arrest Yeltsin or prevent him from making his way to the centre of Moscow. Nor did the army impose a curfew or control movements. It did little to stop half a million protestors taking to the streets of Moscow and soldiers merely looked on as Yeltsin climbed atop a tank to address the crowd. Behind him, on the tank, Yeltsin's allies raised a Russian tricolour. This was no longer a contest between reformist and conservative Soviets; it was a struggle between the Soviet Union and a new Russia. Unwilling to turn their guns on fellow Russians, soldiers refused to clear the streets of protestors. Unable to impose itself, the coup collapsed after just three days. But it was not Gorbachev who had saved the people from the hardliners but Yeltsin. Not Soviet reform, but Russian nationalism. Gorbachev's marginalization laid bare, Yeltsin exacted revenge by making the Soviet president read out a list of conspirators on national television. They were all men close to or appointed by Gorbachev himself. Shortly after, Yeltsin's Russian SSR recognized the independence of Latvia, Lithuania and Estonia.

As Serhii Plokhy demonstrates in *The Last Empire*, in the weeks and months that followed, Yeltsin transferred Soviet authority to the Russian state in a bid to both destroy Gorbachev and replace the Soviet state with a new arrangement led by Russia. It was this that pushed the union's non-Russian republics in the other direction. Forced to choose between a Russian-led "commonwealth" and full independence, they chose the latter. The most crucial vote of all was in Ukraine, which on 1 December 1991 voted by a margin of 92 per cent to 8 per cent to leave the Soviet Union and reject a Russian-led commonwealth in favour of independence. No one – not even Ukrainian leader Leonid Kravchuk – had expected a result like that. Without Ukraine, both the old Soviet and Yeltsin's new Russian projects lacked viability and one by one the remaining Soviet republics followed suit.

It was Yeltsin that pulled the final trigger on the USSR but in truth its fate had been sealed by the Ukrainian referendum. A week later, the president of the Belarussian SSR, Stanislav Shushkevich, worried about the supply of oil and gas needed to keep heaters running that winter, invited the two people who could make things happen, Russia's president Boris Yeltsin and Ukraine's president Leonid Kravchuk, to a hunting lodge in Belavezha to discuss the problem. Yeltsin was eager to finish off Gorbachev, Kravchuk the communist *nomenklatura* to reposition himself as a nationalist. Yeltsin asked whether they should dissolve the

Soviet Union. Since the republics they represented had first constituted the Union, he felt they alone had the right to dissolve it. The others readily agreed. Without consulting Gorbachev, they issued a declaration dissolving the Soviet Union. A leader left without a state, Gorbachev announced his resignation live on air on 25 December. Without ceremony, the Soviet flag was taken down from the Kremlin and replaced by the tricolour of the Russian Federation.

Yeltsin

Russia inherited the Soviet Union's many problems. Its debt. Its dysfunctional economy. Its decrepit industry. Its over-sized military. Its fattened and Sovietized bureaucracy. There was no liberal elite, capitalist tradition, or deep-seated loyalty to the law standing in the background. Everything that existed on the last day of the Soviet Union was still there on the first day of the Russian Federation. Although Yeltsin had been the most vocal champion of reform he had no clear idea about what to do.

The new president prescribed a dose of "shock therapy", a rapid transition to a market economy. Most prices were liberalized overnight causing a spiral of inflation that wiped out savings and pensions. Professional salaries evaporated. Privatization was rushed in the hope that one giant leap would transform Russia's economy. To avoid the self-privatization of firms engineered by middle-managers under Gorbachev, every Russian was given vouchers they could use to purchase shares in privatizing firms. Initially worth $25, the currency collapse meant vouchers were soon worth only $2. Most people had little understanding of what a share was or what it was worth. What they did know was that their savings had disappeared and necessities were painfully expensive. Those who did understand their value was the army of managers who had kept the old system going. They bought up their workers' vouchers at a fraction of their real value and as a result most vouchers and with them ownership of most old firms was simply transferred from public ownership to the private ownership of the *nomenklatura*. Around three-quarters of large firms were bought by the management. Industries were auctioned off at a fraction of their true value. One estimate suggested that the total voucher value of Russian industry was set at a mere $12 billion. Privatization thus achieved only the transfer of wealth from the state sector to the private sector. An uncountable number of other scams saw new capitalists grow very rich, very quickly. They did so not by producing and selling goods and services that people wanted, but by appropriating the state's wealth for themselves. Privatization generated no new investment in Russian industry. Much more money flowed out of Russia than in.

The economic catastrophe was immense. GDP per capita shrunk by about a quarter between 1988 and 1993. That figure masks the true extent of the decline because it includes wealth which continued to flow from the sale of oil and gas. It also masks sharp differences between the Moscow elite and elsewhere. The International Labour Organization estimated that 85 per cent of Russians were reduced to poverty. In some parts of the country, average monthly wages were as low as $6. More than half the country's firms did not pay regular wages. Local governments refused to pass tax revenues on to Moscow. People did what they could to cope. Organized crime, street violence and corruption escalated, fuelled by a mass of under- and unpaid bureaucrats, policemen and soldiers.

Yeltsin soon found himself caught in the same vice that had gripped Gorbachev. This time, the communists – who still controlled the People's Congress – and nationalists joined forces against the president. This "Red–Brown" coalition railed against economic reform and what they saw as Russia's "national humiliation" – their great nation reduced to begging and thievery by criminal and incompetent leaders beholden to the West. Parliament blocked Yeltsin's decrees wherever it could and tried to limit presidential power. At stake was not just the relative authority of president and parliament but the future direction of Russia itself. Yeltsin wanted liberalization, parliament wanted to wind it back. When parliament tried to impeach Yeltsin in April 1993, Yeltsin fought back with a referendum in which Russians showed they still supported the president and his reforms.

Buoyed by his referendum win, Yeltsin issued a decree dissolving parliament. That violated Russia's constitution and drew a predictably sharp response from parliament which declared the decree null and void. Parliament declared Yeltsin sacked and anointed Alexander Rutskoy interim president. A celebrated hero of the Afghanistan war, Rutskoy had served Yeltsin as vice-president. Now, Russia had two presidents and Rutskoy called people onto the streets to defend parliament against Yeltsin – the man who had defended it just two years before. Thousands heeded the call, forming a ring around the parliament building, around which security forces established a cordon. After a tense standoff, several hundred armed activists broke through the police lines to join protestors at the White House, Russia's parliament building. Sensing victory, an exuberant Rutskoy urged protestors to storm the mayor's office and then take control of the television tower at Ostankino. There, parliament's armed volunteers, a combustible mix of communists and nationalists, fought a pitched battle with interior ministry and special forces which left nearly 50 people dead but the security forces still in control of Russia's television. Violence at Ostankino stunned the military leadership and brought it into line behind Yeltsin. The president ordered them to retake the White House and on 4 October, tanks opened fire on the upper floors of the building before soldiers stormed in. More than 100 people were killed as the army took control of the building. Rutskoy was imprisoned but granted amnesty soon after.

Most Russians supported Yeltsin and blamed Rutskoy and his allies for the violence. Nevertheless, the 1993 crisis crushed what little faith they still had in liberal reform. Yeltsin may have been preferable to the nationalist and communist alternatives, but a president willing to unconstitutionally dissolve parliament and then to send tanks against it was clearly no liberal. Disillusion was fed by the collapsing economy and Russians' sense of collapsing pride. The humiliation felt so keenly by nationalists and communists sometimes exacerbated by the president's own actions. A visibly drunk Yeltsin attempted to conduct a band during an official visit to Germany in 1994. Later the same year, he was so drunk during a brief visit to Ireland that he failed to disembark from his plane. What the president needed, the men around him advised, was a small victorious war to restore his political fortunes. Their eyes turned to Chechnya. Lancing Russia's deep wound in the southern Caucuses would hand Yeltsin a victory that could unite the people and consolidate his position. Or so they thought.

Chechnya's campaign for independence had looked much like the others to begin with. It was secular, nationalist and led by established elites. It was the bloody war unleashed by Yeltsin in 1994 that gave it the altogether different hue that it has today. Part of an ASSR when the Soviet Union collapsed, Chechnya had been forced into the Russian empire by Peter the Great, but only after its stiff resistance had been broken. In 1944, Stalin deported close to half the entire Chechen population to the barren steppe of Kazakhstan. Around half perished before Khrushchev allowed them home in 1956. *Perestroika* allowed Chechens to talk openly about this past and organize themselves. An immensely popular leader emerged in the form of Dzhokhar Dudayev. Dudayev was an unusual radical. A strategic bomber commander, he was the first Chechen to reach the rank of General in the Soviet military. Born in Kazakhstan, he was married to a Russian and spoke better Russian than Chechen. Dudayev was killed when a Russian rocket followed the signal from his mobile phone in 1996. By then, however, Dudayev's Chechnya had all-but defeated the Russian army and won itself de facto independence.

When Gorbachev's reforms opened space for political agitation, Dudayev joined a national congress of Chechens. Elections allowed more voices into the Supreme Soviet, including those of Chechen nationalists. Concerned to maintain its popularity, the Soviet was persuaded by the congress to declare Chechnya-Ingushetia ASSR a sovereign state in November 1990. Dudayev supported Yeltsin during August 1991 and when the coup collapsed, so too did Soviet authority in Chechnya. Dudayev and the national congress filled the void, something Yeltsin initially welcomed. But in November, Dudayev overreached by setting a course for Chechen independence from Russia – part of a train of local leaders reacting against Yeltsin's usurpation of Soviet power. Yeltsin tried unsuccessfully to impose a state of emergency. Russian airborne troops tried storming Grozny's airport but withdrew when confronted by armed Chechens. Soon after, Dudayev was elected president and proclaimed Chechnya's independence. Another Russian ASSR, Tatarstan, also declared independence from Russia but in February 1994 agreed to relinquish that in return for substantial autonomy. There followed a two-year pause in Chechnya too where perhaps a Tatarstan-like arrangement could have been found but Yeltsin, distracted by the collapsing economy and his conflict with the Russian parliament, seemed to ignore Chechnya. In the lull, Dudayev missed a golden opportunity to build a functioning Chechen polity and negotiate terms with Moscow. Instead, the republic descended into a violent contest between rival elites funded by spiralling organized crime. A bounty of abandoned Soviet arms meant no one lacked for guns, while geography allowed Chechen criminals to profit immensely from illicit trade into Russia. In a desperate bid to get a grip on the situation, Dudayev dissolved parliament and drifted towards authoritarianism.

Moscow regained its interest in Chechyna in 1994. It provided military support to one of Dudayev's rivals, Umar Avturkhanov, who tried and failed to take Grozny in November. Dozens of Russians working covertly with Avturkhanov were taken prisoner. Then, hoping for a short and victorious nation-building war, Yeltsin ordered the Russian army itself into Chechnya.

A Russian force of 40,000 invaded but took close to three months to seize Grozny. After their initial attack was repulsed with heavy losses, Russian forces attempted to encircle the city and bombard it into submission. Weeks of indiscriminate artillery and air bombardment wreaked terrible vengeance on Grozny's civilians. Some Russian sources suggest more than 20,000 civilians, including 5,000 children, died in those few weeks alone. The city crumbled and Russian forces seized control of the smouldering ruins in March 1995.

But the Chechens weren't defeated. Chechen fighters retreated to the mountains south of the city from where they conducted a highly effective insurgency. Russian losses mounted. Official figures suggest the Russian army lost more than 5,500 soldiers killed and missing and more than 25,000 injured. They responded with another form of atrocity warfare, *zachistka* ("cleansing"). In a typical *zachistka* operation, Russian forces would surround and seal off a village so no one could escape and then proceed to round up the men and boys. Some were shot on the spot, their bodies left where they lay or dumped into mass graves. Others were taken to "filtration camps" and tortured for information. Many of these were tortured to death, many others shot. Some were ransomed back to their families. Sometimes families had to pay to recover the bodies of their kin. Rape and looting were common counterparts. The massacre of more than 100 civilians during one *zachitska* operation, on the village of Samashki, was unusually well reported and stirred global attention. But what happened in Samashki happened in dozens of other places. Chechen civilian losses are difficult to estimate because so little was recorded but estimates put the figure at between 50,000 and 100,000, between 5–10 per cent of Chechnya's entire population. But even this didn't work.

Chechen tactics became more daring and desperate. Shamil Basayev was one commander who rose to prominence. His past clouded in myth, Basayev had allegedly fought with pro-Yeltsin forces during the August 1991 coup. He had then hijacked an Aeroflot plane en route to Turkey demanding Chechen independence, before moving to Abkhazia where he established an armed group supporting the Moscow-backed separatists. Chechnya's war brought him back and Dudayev appointed him as a frontline commander. In June 1995, Basayav led a group of Chechens into the southern Russian town of Budyonnovsk where they seized the hospital and threatened to destroy it unless Russia withdrew from Chechnya. Russian special forces tried several times to storm the hospital, with no success. More than 120 hostages were killed. Finally, Yeltsin agreed to suspend military operations in Chechnya and negotiate. Basayev and his surviving fighters returned to Chechnya taking some hostages with them as human shields.

Things got worse still for the Russians in August 1996. A few thousand Chechen fighters secretly infiltrated themselves into Grozny and conducted a devastating surprise attack. Russian forces stationed in Grozny were routed. Troops stationed outside were surrounded. Decisively defeated, Russia's General Alexander Lebed decided to withdraw his forces and the two sides agreed to negotiate a political settlement by 2001. In the meantime, the Chechens would govern themselves and the Organization for Security Cooperation in Europe (OSCE) would keep a watchful eye on the temporary peace. The first Chechen war was thus another disaster for Yeltsin, the defeat of the mighty Russian army by a tiny separatist group a painful national humiliation. More than anything else, Chechnya 1996 showed just how far Russia had fallen.

Politically and almost literally, Yeltsin was a dead man walking in 1996. His alcoholism frequently on public display, the president was further incapacitated by a series of heart attacks. There seemed no way he would beat communist candidate, Gennady Zyuganov, at the forthcoming presidential election. Even the far-right candidate, Vladimir Zhirinovsky, a crank who campaigned for a return to autocracy, the mass expulsion of foreigners, and the outlawing of homosexuality, outpolled Yeltsin, whose approval ratings plummeted into single digits. But there was no other liberal who could beat Zyuganov and a victory for communism would mean an end to reform. It would have to be Yeltsin. Luckily for the president, Russia's richest people needed him to win and, desperate for that result, Western governments too were prepared to overlook precisely *how* the president clung on to power.

Yeltsin's government struck a deal with Russia's leading capitalists, those who had become extremely rich thanks to the privatization fire sale. The purpose of the bargain was to create a class of extremely wealthy capitalists loyal to Yeltsin's government and capable of securing his re-election. In return for money and support, the government would award them cut-price ownership of Russia's highly profitable strategic industries, its prized oil, gas, coal and nickel. The process was managed through auctions overseen by Russia's new private banks. These banks held government tax revenues which they then used to provide loans to Yeltsin supporters – often the owners of the banks themselves – with which to purchase the industries. The auctions were rigged to ensure that the right people, not necessarily the highest bidders, won. This was how Mikhail Khodorkovsky secured a 78 per cent stake in Yukos, a company worth $5 billion for a little over $300 million. It was also how Boris Berezovsky – owner of Russia's Channel One television station – and Roman Abramovich bought the oil company Sibneft, worth $3 billion, for around $100 million. Berezovsky's Channel One rallied to Yeltsin's cause, as did Vladimir Gusinsky's more independent-minded NTV, which received a huge cash investment from the state-owned resource giant Gazprom. Russia's leading capitalists became the oligarchs. Not just immensely wealthy, but so politically powerful that the government was beholden to them. The oligarchs spent big on getting Yeltsin re-elected. For the rest, the authorities stuffed ballots and stole votes. Yeltsin won re-election, although journalists found it difficult to find anywhere outside Moscow where most people had actually voted for him.

The oligarchs called the shots in Yeltsin's second term. The president suffered a fifth heart attack during the election campaign, which his team had done their best to cover up and he was now very much a back-seat driver at best. Yet things did start to pick up. The economy stabilized as inflation, still high, fell to more manageable levels. Imbued with market ideology, the oligarchs set about restructuring Russian industry. Khodorkovsky's Yukos upgraded its infrastructure and produced yields comparable with traditionally more efficient operations in the Middle East and US. But just as the economy was picking up, catastrophe struck once again. The Asian Financial Crisis triggered a run on Russia's banks. The ruble collapsed (again) demolishing whatever savings people had scratched together since 1992. The government defaulted on its international loans. Three banks went bust. Agricultural subsidies all but disappeared, driving food producers to the wall and causing shortages. Strikes erupted over unpaid wages. The former superpower was bankrupt. Nearly 600 Russians froze to death on the streets of Moscow that winter.

Minds in the Kremlin turned to the question of succession. Yeltsin had long favoured Boris Nemtsov. Urbane, erudite, Western in outlook, he had proven a successful reformer at the city level in Nizhny Novgorod and could be relied on to carry the torch for liberal reform. But Nemtsov was no friend to the oligarchs. In fact, it was Nemtsov that first coined the label "oligarch" to describe Russia's super-rich and politically influential elite. Nemtsov wanted to rein them in, the oligarchs were determined that they wouldn't be. Gusinsky and Berezovsky turned their media empires against Nemtsov and trashed his reputation. The economic crisis did the rest. Market liberalism was a lost cause.

So Yeltsin decided a firm hand was needed. Not a liberal, but someone who could project stability and bring rival factions to heal. Someone with a background in the security services. In truth, Yeltsin's government had never been stuffed with Western-leaning liberals. The government machine he inherited was Soviet. Its staff, its mindset, its practices didn't change overnight with the change of state. The people with most experience of governing and the group best represented in new Russia's halls of power were the *silovki*, the men of the security services, the KGB (now FSB) – which had changed its name but little else. After the White House crisis, one-third of top government positions went to former KGB men. By the end of Yeltsin's first term, that figure was close to a half. First Yeltsin turned to the erstwhile Viktor Chernomyrdin to lead his government and maybe succeed him. Chernomyrdin was relieved of the prime ministership in March 1998 amidst fears he was plotting to seize the presidency for himself. When it transpired, he wasn't, Yeltsin tried to reappoint him but was thwarted by a parliament worried he was too close to the president.

Next, Yeltsin turned to Yevgeny Primakov, a former head of the KGB and foreign minister, someone the communists and nationalists in parliament could not reject. Primakov proved effective at managing the economic crisis. He also burnished his nationalist credentials by taking a strong stand against NATO's 1999 intervention in Kosovo. For those reasons, he was popular too. But that was a problem as he threatened oligarchic interests. Yeltsin was also unsure about Primakov. He was parliament's man, not Yeltsin's. The president wanted a successor who could protect him and his family from future prosecution for corruption when he stepped down. There was one man in the Kremlin, however, who had proven unfailingly loyal. One of the *silovki*, but also a grey and malleable bureaucrat. His name was Vladimir Putin.

Further reading

Svetlana Alexievich, *Secondhand Time: The Last of the Soviets*. New York: Random House, 2017.
Roy Medvedev, *Post-Soviet Russia: A Journey Through the Yeltsin Era*. Translated by George Shriver. New York: Columbia University Press, 2000.
Serhii Plokhy, *The Last Empire: The Final Days of the Soviet Union*. London: Oneworld, 2014.
Vladislav M. Zubok, *Collapse: The Fall of the Soviet Union*. New Haven, CT: Yale University Press, 2021.
Christoph Zurcher, *The Post-Soviet Wars: Rebellion, Ethnic Conflict, and Nationhood in the Caucuses*. New York: New York University Press, 2007.

2
Chechnya

A second war in Chechnya made Vladimir Putin, transforming him from anonymous apparatchik supported by just 2 per cent of Russians into an immensely popular president. It allowed him to renegotiate the state's social contract with the people as the media, oligarchs, regional governments and political opposition were brought under the Kremlin's sway in return for the promise of stability. Chechnya convinced Putin and his allies that war could solve political problems. It taught them the power of nationalist mobilization and demonstrated the capacity of military force to reassert Russia's lost authority and re-establish a sense of pride. It reaffirmed the Russian military's belief in the offensive value of overwhelming and indiscriminate firepower. But whilst the rallying around the flag – and the president – was real enough, some effects were mere mirages. Victory in Chechnya was not as decisive as it seemed. Nor was it achieved by Russian military prowess alone. It was Chechnya's Kadyrov clan that finally won the Chechen war, not the Russian army. Chechnya today may be firmly a part of the Russian Federation but it is Ramzan Kadyrov's word that is law there. His is an autonomous Chechen regime – the sort of thing the separatists wanted – propped up by eye-watering amounts of Russian money.

Operation successor

Vladimir Putin left Dresden at the beginning of 1990, apparently gifted an old washing machine by East Germans grateful for his service. Still on the KGB's books as a member of the active reserve, he returned to Leningrad's university where he renewed his friendship with his old professor, Anatoly Sobchak. The city he returned to was very different to the dormant but safe one he had left a few years before. It wasn't just that its name had reverted back to Saint Petersburg. Gorbachev's reforms had brought freedom and a lawless capitalism. Many of the city's first entrepreneurs were violent criminals. As the economy crashed, food became a prized commodity. Hunger was common. Into the void strode criminal gangs, peddling the essentials of life as well as drugs, alcohol, corruption, prostitution and violence. The old Soviet command system was dead but there weren't yet any laws or institutions to govern the new capitalist system. As the weapons of the Red Army followed demobilized soldiers into the marketplace, AK-47s were readily available from car-boots and street stalls. The old morality was dead too, and it was not yet clear what the new morality should be – much less why anyone should abide by it. Capitalism, many Russians thought, meant every person for themselves in a no-holds-barred struggle for survival and wealth. The country's first capitalists bribed, stole and murdered their way to rapid fortunes. The police disintegrated, unable to match the will, weapons, or wages offered by the criminal gangs. Saint Petersburg was among the worst places; a violent robber's den where even the gravediggers were part of an extortion racket. Here, politics was war.

Putin's sojourn in the academic wilderness was short-lived but much mythologized. He claimed he moonlighted as a taxi driver to earn extra income. Many did do that, but whether Putin did is uncertain. He was soon lifted from the ivory tower by his former mentor, Anatoly Sobchak, recently elected mayor of Saint Petersburg. Sobchak had a reputation (not altogether deserved) as a prominent democrat, a man firmly in the mould of Boris Yeltsin. Beset by the same misfortunes that dogged Yeltsin in Moscow, Sobchak adopted a similarly ambivalent position on democracy and the rule of law. Putin was one of many former KGB agents brought into his administration. Their job, to steady the ship by outplaying the criminals at their own game. It was a game at which Putin excelled. His unstinting loyalty to his boss and unflinching capacity to get things done saw him rise to the position of deputy mayor.

Getting things done in this context meant blurring the boundaries between government and criminality. In 1991, Putin devised a scheme to alleviate Saint Petersburg's food crisis by swapping raw materials for imported food. Materials worth more than $120 million were signed away, with Putin orchestrating the exchange. But the food never arrived, and the money disappeared. Sobchak protected Putin when the scandal was exposed, but he couldn't protect himself from the electorate. The food scandal – one of many to hit Sobchak's administration – discredited the democrats and they were voted from office in 1996, the year oligarchs and electoral fraud helped Yeltsin stumble over the line in the presidential election. The following year, it was Putin that saved Sobchak. As the legal net tightened around the former mayor, his former fixer arranged a private jet to spirit him out of the country without a passport and into comfortable retirement in Paris. Loyalty like that got you noticed by a Kremlin desperate for friends.

Putin was called to Moscow to become deputy head of presidential property. The role sounds arcane, but since the state owned a vast amount of property, the portfolio was a powerful and lucrative one. His loyalty and knack of getting things done for "the family", as the Yeltsins and their closest allies were known, saw him rise quietly through the ranks until, in 1998, he was awarded his dream job – directorship of the Federal Security Service (FSB). Putin led a ruthless campaign of internal reform to drive out any whose loyalty to the government was uncertain, including those who thought fidelity to the law more important than fidelity to their political masters. One of those was Alexander Litvinenko, an idealistic FSB officer who approached his new boss with evidence of corruption in the ranks and a plot to assassinate Boris Berezovsky. Putin drove Litvinenko out of the service, put him on trial, and then drove him out of the country. In 2006, Litvinenko was fatally poisoned with polonium-210 by agents of Russia's Federal Protective Service (FSO), Andrey Lugovoy and Dmitry Kovtun.

Putin also provided personal services to the president. In January 1999, he stopped the public prosecutor, Yuri Skuratov, as he prepared corruption charges against Yeltsin's youngest daughter. Putin released a video showing Skuratov in a compromising position with two prostitutes. Skuratov was forced to resign, the charges against Tatyana Dyachenko made to disappear. Now Putin had Yeltsin's trust and favour, he promised to do what nobody else would: grant Yeltsin and his family immunity from prosecution.

Chechnya provided the catalyst for Putin's elevation. The Khasavyurt peace deal agreed by General Lebed in 1996 had paved the way for a broader deal between Yeltsin and Chechnya's president, Aslan Mashkadov, the following year. But that had not led to agreement on a lasting peace. Rather than consolidating its authority, the Chechen republic descended into tribal strife and organized crime. Radical Islamism, sponsored by wealthy interests in the Middle East, grew in influence. This forced Mashkadov into a series of unpopular concessions, including mandating the wearing of the veil and imposing elements of Sharia. By 1999, Mashkadov controlled little beyond his immediate surrounds, whereas radicals like Shamil Basayev – leader of the Budyonnovsk siege – grew in power. Meanwhile, whipped into an anti-Chechen frenzy by the media, Russian public and military opinion had turned firmly against further negotiations. In early August 1999, Basayev and a Saudi jihadist known as Ibn al-Khattab led a few hundred Islamist fighters into neighbouring republic Dagestan. Precisely why and what they hoped to achieve remains a mystery. Most likely, they overestimated their own popularity and underestimated the local authorities and thus believed they could incite an Islamist rebellion. But there may have been more to it. Intriguingly, Berezovsky spoke to Basayev regularly, negotiating for Russian interests he claimed. Mashkadov, however, alleged the oligarch was bankrolling Basayev and Khattab. In a bizarre twist, Berezovsky admitted giving Basayev $2 million to open a concrete factory for the unemployed. Question marks also hang over the ease with which the invaders were despatched by government forces. Perhaps they knew what was coming? Perhaps Basayev was put up to it by Berezovsky, to give the Kremlin *casus belli* to remedy its flagging fortunes? Perhaps, but perhaps not. Given how the last Chechen war had gone, that would have been an incredibly risky play. The idea of war with Chechnya was not popular in August 1999. But it was the attack on Dagestan that prompted Yeltsin to sack prime minister Sergei Stepashin and bring in a new man: Vladimir Putin.

Yeltsin appointed Putin prime minister on 9 August 1999, effectively anointing him as his successor. The oligarchs, especially Boris Berezovsky, rallied to the little-known bureaucrat too. Putin was a loyal fixer, with no independent wealth, and no obvious political ideas. An ideal president who Berezovsky assumed would be beholden to the oligarchs. By demolishing reformers like Boris Nemtsov, resisting old guarders like Chernomyrdin, and whipping up anti-Chechen vitriol through their media empires, the media oligarchs paved the way for Vladimir Putin assuming he would be *their* man in the Kremlin.

The problem for Putin was that he was up against a potentially winning opposition. To the horror of Yeltsin and the oligarchs, the

Fatherland-All Russian alliance of Moscow mayor Yury Luzhkov and former prime minister Yevgeny Primakov offered Russians a viable conservative alternative that promised stability and an end to the anarchy of capitalist reform. It was one thing to buy an election victory for a seasoned political figure like Yeltsin, as in 1996, another thing entirely to do it for a complete unknown like Putin and in the face of a popular alternative. The media moguls helped. Berezovsky set to work rebranding Putin as a patriotic, tea-totalling, all-action hero; the antithesis of Yeltsin. War in Chechnya did the rest.

It took just two weeks for military and local Dagestani forces to repel the Chechens. The new prime minister arrived shortly afterwards to hand out medals and warn of further attacks. He was right to. On 4 September, a bomb ripped through an apartment block in Buynaksk, in Dagestan. More than 60 people were killed and several hundred injured. Five days later, on 9 September, 109 people were killed and more than 200 injured when a bomb detonated in an apartment building in Moscow. Nearly 120 people were killed four days later when a bomb destroyed another Moscow apartment block. A truck bomb outside a Volgodonsk apartment three days later left 17 more people dead. The country erupted in panic, fear and suspicion. The authorities immediately pointed the finger of blame at Chechen terrorists. The Chechens denied their involvement. To this day, no clear evidence of Chechen responsibility has been presented, leading some to suspect that one or more of the bombings may have been perpetrated by the FSB to rouse public support for another war in Chechnya.

The doubts are fuelled by the strange events in Ryazan, about 200 km southeast of Moscow. A couple of days after the Volgodonsk bombing an eagle-eyed resident in Ryazan noticed two suspicious men carrying sacks into the basement. He called the police, who discovered 150 kg of what transpired to be hexogen, an explosive held in stores by the FSB, wired to a detonator and timer. The apartment was quickly evacuated and the following day, FSB spokesman Aleksander Zhdanovich publicly praised the residents for foiling another Chechen terrorist bombing. But that wasn't the end of it. The suspects' car was identified, and the telephone exchange listened in to a suspicious telephone call which, strangely, it traced to an FSB number in Moscow. The suspects were apprehended and identified as FSB agents. A few hours later, FSB director Nikolai Patrushev issued a statement claiming the whole thing had been a training exercise to test the vigilance of civilians and that the substance found in the sacks was harmless.

What happened and why remains shrouded in mystery, but it is quite possible that Russia's security forces were responsible for one or more of the apartment bombings. Perhaps it was all orchestrated. Perhaps the first one or two were perpetrated by Chechens and seeing how the attacks whipped up public support for war, some FSB officers tried to capitalize by perpetrating more bombings. Perhaps none of the bombings were inside jobs but Ryazan was a bungled attempt to stage an apparent attack that would be foiled by the FSB. But if that was so, why use live explosives and detonators and why not inform at least some people in the local security force? Litvinenko concluded that either the FSB was trying to kill hundreds of Russians or it was trying to create public panic for no obvious reason. Alternatively, perhaps all the bombings were by Chechens and this is a story about an incompetent and corrupt security service that sells explosives to its enemy. Many Russians believed the conspiracies. Several who have investigated them have been imprisoned, harassed, or have suffered untimely deaths.

Whatever the truth, the apartment bombings swung Russian popular opinion firmly behind the man promising to solve Russia's Chechnya problem once and for all. When he took office in August, Putin's approval rating stood at 31 per cent. Only 2 per cent of Russians said they would vote for him as president. By November 1999, just three months later, his approval rating was 80 per cent and Putin was fast becoming a sure bet for the presidency. In the interval, Putin had led a new Russian war on Chechnya.

Putin's first war

Fast becoming the Russian everyman and an all-action hero in the defence of the motherland, Putin quickly put his stamp on the Second Chechen War. This contrasted sharply with his predecessor, Sergei Stepashin, who had preferred a cautious response to the Dagestan crisis. In televised remarks, Putin declared Chechnya a "terrorist state", a "bandit enclave" harbouring foreign-backed Islamic extremists, intent on murdering Russians. He promised to track them down no matter where they hid. The bandits would be liquidated without mercy. He would

Map 2.1 Chechnya
Source: iStock/PeterHermesFurian.

even "waste them in the shithouse". Putin handed out hunting knives as Christmas presents to Russian soldiers serving in Chechnya. Days before the presidential election, in March 2000, he flew into Grozny in a Su-27 fighter jet. The contrast between Putin's hands-on war leadership and Yeltsin's bumbling could not have been starker – although in fact Putin deferred to the military, giving it free rein until revelations about atrocities and the Dubrovka theatre siege forced a rethink. Putin's views on Chechnya resonated closely with public opinion. Public terror after the apartment bombings translated into intense anger. Politicians, media commentators and retired soldiers took to the airwaves demanding Chechnya's annihilation. Putin capitalized. He was young, healthy, sober and intensely patriotic. He was strong, and so was Russia. In just three months, Putin was transformed from unknown bureaucrat to national saviour prosecuting a just war and promising a new contract between the people and their state. More than three quarters of those polled in Moscow and Saint Petersburg said they supported the war; around two-thirds of people in Russia's regions. Wiping out Chechens was now good electoral politics.

By December, the war had become the first step of a master plan for Russian renewal. In a manifesto published as he assumed the acting presidency from Yeltsin on New Year's Eve, Putin promised to restore the country to greatness. He would rebuild Russian pride by re-establishing state power. At the heart of Putinism from the start, then, a political project that united state power and national dignity. "Russia is and will remain a great power". Greatness was in its very being. In its size and geography, culture and history, economy and people. Russia could be nothing but great. Yet, Putin explained, this greatness was in peril. Russia was being pushed downwards and risked becoming a second- or even third-order power. It was the task of government, the government of Vladimir Putin, to correct that. "Russia needs a strong state power and must have it". The state, he wrote, was the only guarantor of order, the only force of positive change Russia had ever known. For in Russia, he explained, the collective always dominated the individual. The hand of paternalism had always been the best guide. Progress emanated from the state, not the endeavours of individuals.

The secret to Putin's success lay in how the war was initially fought. Determined to avoid the mistakes of the First Chechen War, the second war began in September with an intense artillery and air bombardment. Conveniently, this strategy corresponded with Russian military doctrine, inherited from the Soviets. The bombardment of Grozny resembled Soviet tactics of the Second World War, not NATO tactics of the 1990s. The Soviet and now Russian way of war emphasized mass: the use of massive firepower to overwhelm the opponent at distance with infantry employed to mop up and hold territory. Expecting only light resistance in 1994, Russian forces had walked half-heartedly into Chechnya's maelstrom and become bogged down, fighting an enemy that wanted to engage them up-close, where firepower counted for less and surprise, deception and speed for more. This time, they intended to fight at distance with indiscriminate force.

The offensive began in late September with a week-long bombardment of the upper Terek region in Chechnya's north to clear the path for a Russian invasion. Cluster bombs hit the village of Elistanzhi on 7 October, killing nearly 50 civilians. A massive force around 120,000 strong moved in the following month. Standing against them in a republic roughly the same size as Wales or a little larger than Connecticut, Chechnya called on 3,000–4,000 fighters, divided across half a dozen different armed groups that sometimes expressed as much animosity towards each other as towards the Russians. Grozny was attacked by air, rocket and missile. Ballistic missiles and cluster bombs hit a mosque, market and maternity hospital on 21 October, killing more than 130 civilians and wounding hundreds more. The Kremlin claimed the market was an arms bazaar and the other buildings housed terrorists. It saw all Grozny as a terrorist camp and proceeded to level the city. Its electricity, gas, telephones and water were severed as Russian forces encircled the city.

Between 5,000 and 8,000 civilians were killed by the bombing, fewer than 1994 only because there were fewer civilians left in the city and those that were, were better prepared for life under siege. Grozny's surrounds suffered the same fate. The village of Alkhan-Yurt, a few miles to the south, was heavily bombed in early November, killing more than 130 civilians. Typically, towns and villages across the republic were surrounded and shelled before soldiers moved in. The bombing forced tens of thousands of Chechens to flee their homes in search of safety. They found Russia's internal borders closed to them, Putin insisting that terrorists concealed themselves amongst the refugees. Indiscriminate firepower cut off their potential retreat across the border into Georgia. Meanwhile, the government forced more than 10,000 ethnic Chechens resident in the Russian republic back into Chechnya, seizing their Russian passports as they did. At the peak of the war, nearly 250,000 Chechens were forced from their homes – at least half of the republic's population at the time. About a third of those returned home before the end of 2001.

Indiscriminate violence prompted Western criticism. US President Bill Clinton got a sharp response from Boris Yeltsin when he raised the issues at the pair's final meeting, in Istanbul that December. Putin was equally dismissive. He emphasized that Russia's war was part of a global struggle against Islamic terrorism and claimed Osama bin

Laden, orchestrator of the US embassy bombings in Nairobi and Dar es-Salaam, was pulling the strings in Chechnya too. There could be no restraint in such a war, he insisted – an argument that resonated more fully with Americans after 9/11. Russia's relationship with the West had been deteriorating for years. The 1993 White House crisis, the First Chechen War, and rampant corruption had long since ended Western faith in Russian reform and put paid to hopes that Russia might be integrated into Europe's security architecture. Europe and the US still wanted a constructive relationship, but by the late 1990s no one expected more than that. Meanwhile, Eastern Europe's new democracies clamoured to join the European institutions that had brought peace and prosperity to the West after the Second World War. They wanted the sort of protection from future Russia revanchism that only full membership of NATO would give them. If history was insufficient explanation as to why, the smoking ruin of Chechnya offered a contemporary reminder. The Czech Republic, Hungary and Poland followed East Germany into NATO in March 1999.

Russia and the West had also clashed over Yugoslavia. Where the West had looked to contain the war that engulfed the collapsing socialist state in 1991, Russia backed the claims of its ally, Serbia. In 1994, NATO imposed a no-fly zone on Bosnia in response to widespread atrocities by the Bosnian Serbs. The following year, it bombed the Bosnian Serb army, after the latter's genocide in Srebrenica. The West and Russia patched up their differences and worked together implementing the 1995 Dayton peace accord, but the conflict erupted again four years later over Kosovo. Desperate to avoid a repeat of the bloodbath in Bosnia, NATO intervened to stop Serbian atrocities. Russia objected – the charge led by Primakov. After 78 days of NATO bombing, that ended in June 1999, Serbia relented and a multinational peace operation was despatched. Russian forces based in Bosnia to police the Dayton agreement tried to steal a march on NATO by driving down to Kosovo's capital, Pristina, and seizing the airport. Quite what they wanted to achieve remains unclear. Perhaps claim a distinct Russian sector, cause trouble, or remind the West to pay attention to Moscow. The US General in charge of NATO's forces, Wesley Clark, ordered that the airfield be forcefully retaken. Thankfully, the cooler heads – British officers in charge on the ground, General Sir Michael Jackson and future pop star James Blunt – prevailed and a peaceful solution was negotiated.

There was no doubt, however, that Russia and the West were heading in different directions.

Although Grozny was still in Chechen hands, Russian television welcomed in the new millennium with a wave of victory-related programmes. Putin was shown celebrating with the soldiers, telling them they were restoring national pride and ending the country's disintegration. Russian renewal begins here, with me, he was saying. By early January 2000, Russian forces had Grozny surrounded and after a few more weeks of siege and bombardment the outnumbered and overwhelmed Chechen forces retreated from Grozny – stumbling into a minefield where they suffered heavy losses. Shamil Basayev lost a foot. Russians had their victory and Putin his. They had prevailed for three main reasons.

First, Russian forces made greater use of indiscriminate air power and artillery to soften urban areas before sending ground troops in. They employed ground forces more cautiously, sending in small probing groups to identify pockets of resistance that could then be targeted from a distance. Second, Russian operations were better planned, and the army seemingly made greater use of professional soldiers than conscripts, although the extent of that switch was exaggerated by a government and media determined to portray the Russian army as greatly reformed. Third, the Kremlin controlled the media much more effectively than it had in 1994. Access to the front was carefully controlled, so the Russian public was told only what the government wanted it to hear. In this, Putin was helped by a supportive media. Even Vladimir Gusinsky's independently minded NTV took a broadly favourable editorial line, although it didn't completely shy away from criticizing Putin or calling attention to the horrors of war.

The popular narrative was one of unremitting success. Putin had delivered Russia from the grip of Chechen terrorists. He had given Russians new belief in themselves, the armed forces and their state. Seeing the writing on the wall, Primakov withdrew from the presidential race. Putin duly won a resounding election victory in March, winning more than 50 per cent of the overall vote, well ahead of the nearly 30 per cent won by the communists' Gennady Zyuganov. In a largely free vote, Putin secured 18 million more votes than his main rival and was the most popular candidate in all but five of the country's 88 federal units.

But victory in Chechnya was more apparent than real. Chechen forces suffered losses as they retreated, but most got out of Grozny. As in 1994, they intended to fight a long insurgency. The war was far from over. Meanwhile, Russia's media underestimated their own side's losses. By January 2000, more than 1,200 Russian soldiers had been killed and 5,000 wounded in Chechnya – a casualty rate comparable to Afghanistan and the First Chechen War.

Long war

After the fall of Grozny, the war in Chechnya became a grindingly brutal insurgency punctuated by a campaign of terrorism that reached an horrific nadir in Beslan.

Chechen insurgents concealed themselves in the countryside and employed improvised explosives (IEDs) and hit-and-run tactics to exact a steady toll on Russian forces. Russian counterinsurgency centred on the *zachistka* – "sweep" or "cleaning" operations whose calling cards were indiscriminate killing, torture, rape and generalized looting. The first recorded *zachistka* of the Second Chechen War came before the fall of Grozny. After intense bombardment, Russian forces entered Alkhan-Iurt on 1 December. Men were separated from women and checked for involvement in the war. Some were shot on the spot. Others were killed with grenades as they huddled in cellars. The town was looted and some who resisted were also shot. Human rights organizations reckon that between 20 and 40 people were killed during the *zachistka*, a pattern that would be repeated many times over. Putin awarded the "Hero of Russia" medal to the commander of the forces involved in this atrocity, General Vladimir Shamonov. Putin announced that military reform was one of his top priorities, a cornerstone of his bid to rebuild the Russian state and, with it, Russian pride. The defence budget was increased by nearly a fifth, from around 3 per cent of GDP in 1999, to nearly 4 per cent by 2002.

The *zachistka* became near ubiquitous. Typically, they lasted between one day and three weeks. A village would be encircled by regular forces (often conscripts) using armoured vehicles, trucks and artillery pieces supported by helicopters to prevent fighters and civilians alike from fleeing. Sometimes, artillery and airpower would be used to flush out fighters, especially if there was any returning of Russian fire. The sweep itself was usually conducted by the FSB, the military intelligence (GRU), special reaction forces (OMON or SOBR), and *kontraktniki* troops hired specifically for the task. These forces moved from house to house. Those they suspected were connected to the insurgency were rounded up and taken for further interrogation. Many were executed immediately, many more were tortured. Many were kidnapped and ransomed back to their families. Sometimes, families were required to pay for the return of their loved ones' bodies.

A *zachistka* was conducted in Novye Aldy, to the south of Grozny, in early February 2000. The area was surrounded, and soldiers moved in. Russian soldiers separated men from women. They looted. They burned down some houses as they went. In others, they wired up grenades as booby traps. They looted and shot at civilians who witnessed their crimes. They also executed their captives as they stood on street corners. At least 56 civilians were killed that day, including children.

Prisoners were often taken to the dilapidated prison at Chernokozovo. The Russians called it a "filtration camp" to detain and evaluate suspected insurgents prior to release or trial. It became a notorious torture centre. Detainees were beaten, electrocuted, held in stress positions for long durations, and subjected to "water-boarding". Sexual violence, including genital mutilation and rape, was common. Most detainees were eventually released, some only after soldiers extorted money from their relatives. Among them was Zura Bitiyeva, who after her release filed a complaint with the European Court of Human Rights. In 2003, Bitiyeva and her family were killed by masked men. The Court ruled that the Russian state was responsible. Other detainees didn't get that far and were summarily executed, typically either shot or thrown from helicopters from heights which killed but not instantly.

Despite the authorities' best efforts, the Chernokozovo torture centre and the *zachistka* killings were widely publicized. Although Russia's media moguls backed Putin, Russia still enjoyed press freedom – a legacy of Gorbachev and Yeltsin. Abuses were documented by the courageous, such as Memorial, a human rights NGO first established to document the grim realities of the Gulag, and journalists like Anna Politkovskaya. The reported abuses dented the government's narrative of a clean, precise and decisive war against Chechen terrorists. Politkovskaya wrote of a new gulag system in the Caucuses. Boris Nemtsov compiled evidence of abuses, which he condemned. Acting commander, Lieutenant General Vladimir Moltenskoy, acknowledged crimes had been committed and introduced new rules purporting to end the torture and civilian killing. Yet arbitrary detention, torture, extortion and murder continued. Chernokozovo was cleaned up and the dirty business of "filtration" moved to temporary sites and torture pits. *Zachistka* killings declined, replaced by an epidemic of "disappearances". According to Memorial, more than 500 people were "disappeared" in 2002 alone. Eighty-one bodies were subsequently found. More than 360 of the abducted were never seen again. By 2006, the number of Chechens killed in *zachistkas* or forcibly disappeared numbered around 8,000.

Still the insurgency ground on. By the end of 2001, Russia had lost 4,000 dead and 13,000 wounded soldiers. Another 500 died in 2002. The death rate declined somewhat in the following two years, but the number of injuries increased. By 2004 Russia was still losing military helicopters to Chechen fire at the rate of about one per month.

This was not the clean war Putin wanted. The events of 9/11 helped by compelling the US to prioritize its own counterterrorism campaign. When George W. Bush announced his "War on Terror", Putin was quick to insist that Russia was already at war with radical Islamism, in Chechnya. That muted international criticism, but it couldn't win the war. And in 2003, the Chechens opened a dangerous new front, launching a wave of terror attacks that brought the war into the very heart of Moscow and threatened to upend Putin's victory narrative. During Putin's first presidential term, only Afghanistan and Iraq suffered more terrorist attacks than Russia. And this was just the tip of the iceberg of Russia's enduringly violent society. At 29 per 100,000, the country's annual murder rate was higher during Putin's first term in office than it had been under Yeltsin, where it averaged 27. Whatever the official narrative, Putin had not stabilized the country. But he had exhibited power and strength, and that was warmly welcomed even if it sometimes did Russians more harm than good.

The Chechens, of course, had form with terrorism. They had employed it during the first war, most notably Basayev's attack on the Budyennovsk hospital which had extracted concessions from Moscow. The new front was opened in May 2002, when an IED targeted Victory Day celebrations in Kaspiysk, Dagestan, killing 45 and wounding more than 200. In early October, a bomb exploded outside a McDonalds in

Moscow. Fortunately, there were no casualties. The same could not be said of the attack on the Dubrovka Theatre in the city's centre. On 23 October, more than three dozen Chechen terrorists, some suicide bombers strapped to explosives, including "black widows" – the surviving wives of Chechens killed by Russian forces – stormed into the theatre during a show, taking 900 hostages. They demanded that Russian forces withdraw from Chechnya within one week and threatened to kill the hostages if they did not. Putin decided it was more important to convey strength than seek a negotiation. On the third day of the siege, Spetsnaz special forces pumped gas into the theatre to subdue the terrorists and then stormed it. The terrorists were overcome and shot but the special forces had not bothered to coordinate their actions with medical services. One hundred and thirty hostages died, all except two killed by the gas employed to save them.

The show of strength did not subdue Chechen terrorism. In 2003 alone, there were nine suicide bombings in Moscow and more than 600 across the country. In July, Chechen suicide bombers killed 15 and injured 60 at a rock music festival at Tushino airfield. In December, four Chechens killed 46 and injured more than 170 when they detonated bombs on a commuter train in Stavropol Krai in the northern Caucasus. In February 2004, suicide bombers targeted the Moscow metro, killing another 41 and injuring 120. A market in the city of Samara, on the Volga River, was targeted in June, 11 killed. Three bus stops in Voronezh were bombed the following month. On 24 August, suicide bombers boarded two domestic flights at Moscow's Domodedovo airport. Both planes were blown up and 90 people killed. The Moscow metro was bombed again the same month, and ten more people lost their lives. Another 50 were wounded. The country was in crisis. Putin's leadership should have been in question.

That it wasn't was due in large part to the fact that the new man in the Kremlin had already engineered a media revolution. Putin had risen to the top partly because Yeltsin's oligarchs thought they could control him. Vladimir Gusinsky had already fallen foul of the Kremlin thanks to NTV's critical coverage of Chechnya. Putin viewed criticism of the war as tantamount to treason. He held the media responsible for the disastrous First Chechen War since its criticism of the war had eroded popular support and military morale. Now, he perceived the independent media as a fifth column, threatening to undermine the state. But whereas Yeltsin, for all his faults, was instinctively liberal and thus did little to curtail press freedom, Putin had no such qualms. In his first year in office, he brought all the country's major television stations under state control. Gusinsky was first. Tax charges were filed against him in May 2000, and the following April, NTV was brought under effective Kremlin control when the state energy giant Gazprom became its main shareholder. Gazprom had loaned money to help Gusinsky buy NTV as part of the loans-for-shares deal engineered by the "family". Facing corruption charges, Gusinsky was arrested but then fled into exile. Gazprom grabbed his NTV shares.

Berezovsky was next. Putin's first major crisis came not in Chechnya but in August 2000 when the nuclear submarine *Kursk*, Russia's most advanced, sank in the Barents Sea following an explosion onboard. Some crew members survived the initial blast, but the Russian navy rejected US and European offers of help with their rescue although it couldn't free the sailors itself. Putin didn't even break from his holiday on the Black Sea. All 118 sailors were lost, killed either by the accident or by asphyxiation days later. In response, the Kremlin orchestrated a badly conceived cover-up reminiscent of Soviet days. Putin badly mishandled a meeting with mothers of the dead. All this was exposed on Russian television and so Putin went after other media moguls. They were the source of Russia's weakness, he insisted. They bred corruption, sowed division, undermined faith in the state. Berezovsky was driven out of the country by corruption charges. He held on to Channel One until the end of 2002 when it was forced into state hands. In 2013, Berezovsky was found hanged at his house near London.

The Kremlin's control of the media was not yet what it would later become. NTV, for example, offered relatively independent coverage of the Dubrovka siege. But the national media was well and truly under the cosh by the end of Putin's first year in office. It helped Putin win an even more decisive victory in the March 2004 presidential election, scooping up 72 per cent of the vote and 40 million more votes than his nearest rival, the communist Nikolay Kharitonov. Putin used his media platform to speak of making Russia great again, even as Russians died on the *Kursk*, in the Dubrovka theatre and in Chechnya, and it won him enormous support. It helped too that by 2003 the economy was starting to take off thanks to rising energy prices and economic reform.

Chechen terrorism reached its terrible nadir just a day after the second Moscow metro bombing. The first day of a new school year is a big event in Russia. As families gathered to celebrate at Beslan's Middle School No. 1 in North Ossetia, heavily armed Chechen terrorists, including "black widows", herded 1,300 children, parents and teachers into the gymnasium, which they wired with explosives. They shut off ventilation and water and smashed windows to prevent a repeat of Dubrovka. The hostages, mostly children, endured awful conditions, denied water and access to toilets. The authorities were once again in no mood to negotiate. But neither, for all Putin's talk of military modernization, were they able to respond with effective force. Things came to their tragic end on the third day. An explosion – accidental according to some sources, caused by a Russian round hitting one of the bombs say others – caused some of the hostages to make a dash for freedom. The terrorists opened fire and the security forces waded in. Panic and chaos ensued. Gun battles lasted for hours. The bombs in the gymnasium detonated and the roof collapsed; although other accounts say it was Russian artillery that brought down the roof. Either way, 333 hostages were killed, including 186 children.

Russia was stunned. The attack had exposed serious problems with the security forces but the palpable sense of horror at a massacre of school children vastly outweighed any criticism of the government. Putin was quicker this time to grasp the national mood. He sacked the senior leadership in North Ossetia and then introduced sweeping reform of the Russian Federation itself. Parliament and public acquiesced as Putin scrapped gubernatorial elections. From now on, regional governors, the executives of Russia's federal entities, would be appointed by the Kremlin. This had little to do with countering terrorism – the operation at Beslan has been overseen by the federal FSB after all. It had everything, however, to do with the centralization of political power in Putin's hands. The move crystallized Putin's "power vertical", Russia's new authoritarianism. The power vertical is a system of government based on patronage. Within this system it is not formal institutions that matter but one's relationship to the hierarchy, one maintained by informal rules which everyone understands. Junior people owe loyalty to their patrons in return for which patrons grant them a share of their wealth and status. Individuals climb the vertical by demonstrating loyalty to their patron. Atop the system sat Putin, now empowered by law to appoint every governor in the land and thus in effect, by virtue of the power vertical, to control every appointment to every state position everywhere in Russia. (Since those directly appointed earn their appointment by loyalty to Putin demonstrated by their capacity to ensure that those beneath them on the vertical also demonstrate loyalty).

Kadyrov's kingdom

The collective sense of horror and hatred unleashed by the atrocity at Beslan could not mask the fact that victory was nowhere in sight and war weariness was growing. A new approach was needed. It was found in "Chechenization". This involved transferring responsibility for the war from the federal authorities to Chechens themselves. The Kadyrov clan was identified as a viable partner early in the war but efforts to devolve responsibility accelerated after the Dubrovka siege and became the main effort after Beslan. The clan was led by Ahmad Kadyrov, Chechnya's former mufti. Ahmad had declared holy war on Russia in 1995 but then become disenchanted by the growing influence of foreign Islamists and the movement's turn towards Salafism. The Kremlin promised him money, power and authority, and in return Kadyrov agreed to become its man in Chechnya, appointed head of a Russian-backed administration in mid-2000. Responsibility for the war was gradually transferred from the FSB, GRU and Ministry of Internal Affairs (MVD) to locally recruited Chechen forces.

In return for their service, the Kadyrovs were awarded a freehand to conduct their affairs however they pleased. They received substantial payments from the Kremlin as well as control of Chechnya's formal and more lucrative informal markets. The cash was used to raise Chechen armed units loyal to Moscow, commanded by Ahmad's son, Ramzan, and to buy off other disenchanted clans. Chechenization was dealt a blow when Ahmad was assassinated in 2004, but Ramzan stepped in and proved up to the task. The new lord of Chechnya was an astute choice. Ramzan employed extreme violence (the Kadyrovtsy mastered the art of the disappearances as well as torture and killing) and bribery to persuade Chechen clans to abandon the insurgency. In this, he was helped by the killing of Mashkadov in 2005 (likely at the hands of the FSB) and Basayev the following year (likely an explosives accident, but the FSB claimed responsibility), which deprived the insurgency of its leaders. Beslan, meanwhile, undermined many Chechens' sense of legitimacy.

It was the Kadyrovtsy that defeated the Chechen insurgency, not Putin. Handed absolute power in his personal fiefdom, Kadyrov rules arbitrarily unconstrained by Moscow. Article 1 of Chechnya's constitution declares the republic to be "independent". Its sovereignty "indivisible". Chechnya, it says, determines its own internal and external policy. It is an independent state in all but name, almost entirely paid for by Moscow. In 2010, Moscow was subsidizing Chechnya at the rate of $1,000 per head, more than six times the national average. Ninety per cent of Kadyrov's budget came directly from Moscow. Whatever cash he could extract out of Chechens he kept for himself. With Russian money, Grozny was rebuilt, a Potemkin city of glass skyscrapers, boasting a $280 million football stadium to host Kadyrov's team and a mosque to rival the Hagia Sophia. The president built himself a shimmering palace, complete with private zoo. He invited Western celebrities to perform at his thirty-fifth birthday party. Hilary Swank, Jean-Claude Van Damme, Vanessa Mae and Seal performed for Kadyrov. All this funded by the Russian taxpayer in return for a stability resting on the shaky foundations of state terror and corruption.

Anvil of Putinism

Putin's presidency was forged on the anvil of Chechnya. Fear of Chechen terrorism shaped public demand for a new type of leadership. Putinism both fed and satisfied that demand. The new president was everything the old president was not. He projected strength. No compromise. Terrorists killed in the shithouses. He was sober, fit and healthy. He joined the troops on the front lines. He flew in fighter jets. The war in Chechnya didn't just propel Putin to power by transforming a grey bureaucrat into a Russian everyman superhero, it laid the foundations of a new social contract between people and state. Putin would give the people stability, security and pride in themselves. He would do this by the centralization and strengthening of state power, the remilitarization of Russian society, and the restoration of national pride through nationalism. All that was asked in return was that the people place their faith in their president and give him their allegiance. It was a contract most Russians were happy to accept. Putinism was thus a partnership between state and people, its creeping authoritarianism simultaneously engineered from the top-down and bottom-up.

But in truth victory in Chechnya was costly, contingent and ultimately achieved by Russia's money not its military might. The mythos of Putinism forged in Chechnya had something of the Potemkin about it. It was the Kadyrovtsy that ultimately subdued Chechnya and their price for that was the establishment of a self-governing and sovereign kleptocracy ruled independently of Moscow but bankrolled by vast amounts of Russian money. To achieve that,, Russia lost more than 7,500 soldiers killed and close to 30,000 injured in Putin's first war.

Further reading

Emma Gilligan, *Terror in Chechnya: Russia and the Tragedy of Civilians in War*. Princeton, NJ: Princeton University Press, 2009.

James Hughes, *Chechnya: From Nationalism to Jihad*. University Park, PA: Pennsylvania State University Press, 2011.

Arkady Ostrovsky, *The Invention of Russia: The Journey from Gorbachev's Freedom to Putin's War*. London: Atlantic, 2018.

Anna Politkovskaya, *A Small Corner of Hell: Dispatches from Chechnya*. Chicago, IL: University of Chicago Press, 2007.

Valery Tishkov, *Chechnya: Life in a War-Torn Society*. Berkeley, CA: University of California Press, 2004.

3
Georgia

Putin got his small and victorious war. But it came in Georgia, not Chechnya. Russia's August 2008 war in Georgia proved immensely popular at home. With only a modest display of its military might, Russia seemingly achieved all its political objectives. It is ironic, then, that this triumph arrived only after Putin had stepped down as president.

Russia's war with Georgia remains mired in controversy centred on the question of who fired the first shot. The Georgians insist Russia conducted a campaign of destabilization aimed at a creeping annexation of Georgian territories in South Ossetia and Abkhazia. This campaign had its roots in the politics of Soviet collapse but accelerated under Putin. They argue that in the spring and summer of 2008, Russian-backed South Ossetian separatists attacked Georgian villages and security forces and that Russia deployed troops into the disputed territories to support them. Faced with the actualization of creeping annexation, they say, Georgia had little choice but to send its own troops into South Ossetia to hold the country together. The Kremlin claims it intervened to protect the South Ossetians from genocide only after Georgian forces attacked.

To understand what happened and why we need to trace two stories that conjoined. The first is a Georgian story: of its tumultuous separation from the Soviet Union, civil war and authoritarian government. The second is a Russian story: of how the Kremlin attempted to preserve what it called a "zone of privileged interest" in the former-Soviet space, a "libcral empire" to use the terminology employed by Anatoly Chubais, arguably the chief architect of the Yeltsin government's economic reform. Russia's first efforts to control the newly independent governments on its borders were chaotic, but over time they morphed into something more coherent and substantial, a core of Putin's political project. The two stories conjoined in 2003 when Georgia's "Rose Revolution" upended the rule of Eduard Shevardnadze (Gorbachev's long-serving foreign minister turned president of Georgia) and replaced it with that of Mikheil Saakashvili: a democrat (initially), reformist and nationalist. Saakashvili wanted to do two things: drag Georgia out of Russia's sphere and into closer partnership with the West and establish the state's authority over all the country's territory. That put Saakashvili on a collision course with Putin. At stake was the question of whether Georgians or Moscow would determine Georgia's political identity and international allegiances; of whether Georgia was a properly sovereign state, or a semi-sovereign entity still in thrall to Russian imperial power. That question was not confined to Georgia. Ukrainians posed it too just one year after the Rose Revolution.

Georgia: civil wars, national collapse

Georgia's separation from the Soviet Union was unhappy and bloody. The country was already dissolving when nationalist leader, Zviad Gamsakhurdia, was elected its first president a month after its declaration of independence in April 1991. The Georgian economy, once among the Soviet Union's wealthiest, was in freefall. Its politics erratic. Gamsakhurdia, the nominal democrat, refused to denounce the August 1991 coup attempt or pledge support for Yeltsin. That was noticed by a new Russian government renowned for bearing grudges. Meanwhile, Abkhazia, South Ossetia and Adjara had already broken away from Tbilisi.

In January 1991, Gamsakhurdia despatched his new national guard to seize back control of Tskhinvali. But this ragtag new force was better at looting than policing and was quickly withdrawn after meeting fierce resistance from South Ossetian separatists. Georgian militia looted and burnt as they went, Tskhinvali was heavily damaged. But when they encountered resistance some national guard units simply refused to fight. They tried again in March, April and September that year but with the same result each time. Unable to hold the town, the Georgians blockaded it instead. Electricity and gas severed, Tskhinvali was plunged into a terrible and frozen winter. By now, however, the wider conflict was taking form. The separatists held the town, the Georgians the mountains around it. Beyond Tskhinvali, where one-third of South Ossetia was inhabited by Georgians, the land became a patchwork of Georgian and Ossetian held villages with no clear front lines. Around one thousand people were killed to produce this stalemate.

Georgian politics fell into chaos. In December, disgruntled members of Gamsakhurdia's inner circle, supported by Soviet troops, launched a coup. When the president refused to step down, fighting erupted between the national guard and those still loyal to the president which left more than 100 people dead. Gamsakhurdia fled the capital and moved to his home near Poti in the country's west, just south of Abkhazia from where the "Zviadists" conducted a low intensity insurgency. Back in Tbilisi, the coup leaders brought in Edvard Shevardnadze to lead a new government. Shevardnadze was seen as the ideal antidote to Gamsakhurdia: experienced, pragmatic and inclined towards democracy. Yet the new leader faced an immense challenge to hold his country together.

First there was South Ossetia. Things had settled down since the violence of 1991, in part due to an earthquake in late April, but arms had poured in giving both sides more firepower. Aware his authority over the national guard was tenuous, Shevardnadze wanted to negotiate a deal. Things spiralled out of control, however, in June 1992, when Georgian irregulars massacred 36 South Ossetian refugees as they journeyed towards North Ossetia. Outraged North Ossetians demanded that Russia intervene. Nationalist politicians in Moscow agreed. The Kremlin suspended gas supplies to Georgia. As violence escalated across South Ossetia, there were reports of Russian tanks and helicopters firing on Georgian positions. At this moment of acute crisis, Shevardnadze and Yeltsin met in Dagomys, near Russia's Black Sea resort of Sochi, and cut a deal which effectively also set the terms for Georgia's future relationship with Russia.

The agreement said nothing about South Ossetia's political future but provided for a ceasefire and the demilitarization of conflict zones. Former Soviet forces were to be withdrawn and, in their place, Russian forces would be deployed alongside Georgian (at a ratio of 2:1) as peacekeepers. The arrangement would be monitored by a small UN observation mission (UN Observer Mission in Georgia). Dictated by Moscow, the deal gave Russia a military foothold inside Georgia through which it could influence both Tskhinvali and Tbilisi. It was a step towards bringing Georgia back into Russia's "liberal empire". The Dagomys agreement was concluded on 24 June 1992. Two weeks later, war erupted in Abkhazia.

South Ossetia matters very little to Moscow. It has a tiny non-Russian population of a little over 53,000. It is small, around the size of Majorca or Long Island. Its land is mountainous, it has no resources, and it does not sit in a strategic location. The South Ossetians have ethnic kin in Russian North Ossetia, but they too are far from numerous, numbering only around 750,000. Yeltsin's bargaining was primarily about using South Ossetia to influence Tbilisi. Abkhazia, however, is a different matter entirely. It is wealthy. It has access to the Black Sea. It sits on important oil pipeline routes from Azerbaijan. Russian elites have holiday homes there. But Abkhazia was as complex as it was important. The Abkhaz were a small, privileged minority even within Abkhazia and this explains why there was no mass violence there during the tumult of Soviet collapse.

Two things had changed by August 1992, however. Since the "Zviadists" had set up shop in Poti, Abkhazian separatism became more of a problem for Tbilisi, a potential springboard from which an attempt to retake government might be launched. Meanwhile, the Dagomys agreement had established a template for Abkhaz separatists to follow. Both sides therefore acquired an interest in fighting that they had not had before. On 14 August, the Georgian National Guard attacked Abkhazia's principal town, Sukhumi. It remains unclear whether Shevardnadze ordered the incursion or whether it was orchestrated by the head of the national guard, Tengez Kitovani. A figure who enjoyed close ties to the Russian defence ministry, Kitovani had been an associate of Gamsakhurdia. After the war, he tried (and failed) to oust Shevardnadze and then in 1993 led a ragtag militia on a wild mission to reclaim Abkhazia. He was foiled by Georgian police and imprisoned, after which he was exiled to Russia.

Kitovani's national guard made some progress, but familiar problems soon arose. Guardsmen were disorganized and disorderly. Predictably, violence and looting stirred local opposition. Civilians fled north and many joined the Abkhaz militia. Their numbers were bolstered by around 1,000 volunteers, among them Shamil Basayev, who poured in from the northern Caucuses. Russian forces not yet withdrawn from Soviet bases provided the separatists with arms and sometimes served as guns for hire. Russian sources report that Abkhaz forces were supplied 70 tanks with their Russian crews. A Russian Su-27 fighter jet and its Russian pilot was shot down over Abkhazia in October. It is unclear whether Moscow was orchestrating Russian military involvement or whether this was a case of entrepreneurial soldiers and pilots cashing in on their skills at a moment when they were left unpaid and abandoned by their own state. Moscow also sold weapons to Tbilisi and tried to mediate. In October, Basayev led a dramatic surprise attack on Gagra that swiftly defeated the Georgians based there. The war dragged on through the winter and into summer 1993, evolving into a series of sieges and counter-sieges. Both sides employed ethnic cleansing and the war seemed to stalemate. Moscow mediated a ceasefire and the withdrawal of heavy weaponry in July 1993. Yeltsin also proposed the deployment of Russian peacekeepers on the South Ossetian model. The Georgians refused, fearing this would effectively separate Abkhazia from Georgia. So Russia stepped up military support to the separatists and in late September Abkhaz forces, seemingly backed by the Russians, conducted a devastating offensive which caused Sukhumi to fall quickly. Some Georgian government officials who refused to abandon their posts were shot. Within two weeks, the separatists had achieved control over all Abkhazia. Looting, ethnic cleansing and mass displacement followed; an unmitigated disaster for Georgia. Of the 250,000 Georgians that had called Abkhazia home, only a few thousand now remained.

It was then, in September 1993, that Gamsakhurdia moved against Shevardnadze. "Zviadists" forcibly disarmed some national guard units as they retreated from Abkhazia presaging a move on the capital. With Georgia on the brink of collapse, Shevardnadze finally accepted Yeltsin's terms. In return for peace in Abkhazia and help with the "Zviadists", he agreed to renew the leases on Russian bases in Georgia and to accept the more-or-less permanent deployment of Russian troops in Abkhazia, 3,000 troops deployed as "peacekeepers" under the auspices of the Commonwealth of Independent States (CIS). Most crucially of all for the Kremlin, Shevardnadze agreed that Georgia would join the CIS, the political association cooked up by Yeltsin in the Belavezha forest as a replacement for the Soviet Union, the institutional arrangement through which Moscow intended to manage its "zone of privileged interest". Georgia was also required to join the Collective Security Treaty – intended as Russia's answer to NATO. By joining these bodies, Shevardnadze was effectively agreeing to bind Georgia into Russia's new imperial space in return for conflict freezes in South Ossetia and Abkhazia and help against his domestic opponents. The Kremlin kept

Map 3.1 Georgia

Source: iStock/PeterHermesFurian.

its side of the bargain. Russian marines moved against Gamsakhurdia's base in Poti, and the Georgians were given military aid to defeat the "Zviadists". Gamsakhurdia died in mysterious circumstances in December 1993, most likely by assassination or suicide.

Georgia's troubled early years had thus produced a political bargain with Russia. In return for a stability of sorts, Georgia bound itself to Russia's sphere of influence. It was when Georgia sought to renegotiate that arrangement that it found itself on a pathway to war.

Russkiy mir: Russian world

Russia's road to war in Georgia also began in the years following Soviet collapse. Moscow's first response to losing the Soviet Union was ambivalent. On the one hand, there was imperial fatigue; a sense that Russia too was a victim of Soviet empire and that by cutting away the other republics, Russia might free itself. Imperial fatigue was wrapped up in Yeltsin's struggles, first with Gorbachev and then with the communists in parliament, both political struggles that pitted Russia (Yeltsin) against the Soviets. Fatigue was connected to collapse since even had Russia wanted to claim the Soviet empire, it was in no position to do so. On the other hand, however, imperialism was interwoven into the very fabric of the Russian state, and of what it meant to be Russian. Since Peter the Great at least, modern Russia has defined itself as an exceptional and imperial state, a state whose relationships with neighbours should be based on hierarchy not equality. Thus, Yeltsin's first instinct after August 1991 had been for Russia to supplant as much of the USSR as possible not to fragment it into a multitude of genuinely sovereign states. When Russians found the time to think about their national identity in the wake of Soviet collapse, a vision of Russian imperium was rarely far from the surface. Anatoly Chubais's concept of "liberal empire" tried unsuccessfully to marry this imperial identity to liberal reform.

Moreover, even at its most fatigued, Russia's understanding of how international politics ought to be conducted was very different to the West's. The Western world embraced liberal international order: a society of self-determining and human rights respecting states, which recognize one another's sovereignty, territorial integrity and right to choose their own path. This was a view of the world encapsulated in the "Helsinki Accords", the "Final Act" of the Conference on Security and Cooperation in Europe agreed by 35 states, including the Soviet Union and United States, in 1975. Although it formally endorsed Helsinki, Soviet policy and practice embraced a very different idea of world order. According to this vision, world affairs should be managed by great powers assuming responsibility for affairs within mutually recognized spheres of influence. This was how Peter and Catherine the Great saw the world; it was the vision Stalin articulated at Yalta in 1945; and that the US, UK, France and China seemingly embraced in becoming permanent members of the UN Security Council; it was how the Soviet Union conducted itself through the "Brezhnev Doctrine"; and it was how Russian foreign-policy-makers assumed things would continue after Soviet collapse. A world order based on Yalta, not Helsinki.

The idea that Russia ought to control a sphere of influence, be it labelled an "empire", "liberal empire", or "zone of privileged interests" runs deep. Russian poets, composers, writers, orators and leaders have portrayed it as a matter of national destiny. Moscow as the "Third Rome", Russia defender of the true Orthodox faith. Russian historians and strategists have often portrayed this as necessary to the nation's very survival. Russia needs an empire, their logic runs, to protect it from the persistent threat of invasion, whether by Mongol hordes from the east (as in the fourteenth century) or Nazi hordes from the west (as in the twentieth century). Even at its lowest ebb, the predominant image of Russia portrayed by state and society never fully relinquished these imperial claims. Russia's political leaders and strategic thinkers also tended to assume that this was how everyone else saw the world too. Thus, to their mind, NATO and the EU were not voluntary associations of states cooperating for mutual benefit, but rival spheres of influence. That these institutions proved so successful that former Warsaw Pact members and Soviet republics clamoured to join them seemed only to confirm the idea of great blocs competing for the loyalty of smaller states lacking in power, agency and sovereignty.

Yeltsin's government devised various schemes to institutionalize Russian primacy in the post-Soviet space. The CIS was conceived as a conduit through which to tie former Soviet republics into the Russian political and economic orbit. It was accompanied by the Collective Security Treaty, which evolved into the Collective Security Treaty Organization (CSTO) in 2000, a regional alliance partly modelled on NATO but with the great power at its core wielding influence more overtly. For example, whereas no American has ever been appointed secretary-general of NATO, five of the seven CSTO-heads have been Russian. Other schemes included the 2000 Eurasian Economic Community, which brought Russia together with Belarus, Kazakhstan, Kyrgyzstan and Tajikistan into a European Community-style economic union, supplanted in 2015 by the Eurasian Economic Union (later the Eurasian Union), a single market with some common political institutions including Russia, Belarus, Kazakhstan, Kyrgyzstan and Armenia.

Two states sat at the heart of Russia's imperial imaginings: "White Russia" and "Little Russia" – Belarus and Ukraine. Russia and Belarus agreed to form a "Union State" in 1996, with a single economy, shared elected bicameral parliament, state council, and much else. Three years later, they pledged to adopt a common currency, flag, presidency, citizenship and army. The scheme to reassemble these two pieces of the "Russian world" ran aground as it became clear that while the Kremlin was trying to extend its control over Belarus, Belarus president, Alexander Lukashenko, harboured his own ambition to use the Union State as his pathway to the presidency of a greater Russia. The union floundered between these two poles, but the vision did not altogether disappear. During Putin's first two terms in office (2000–08), Belarus – its economy trashed by corruption and authoritarianism – was made increasingly dependent on Russia. Putin propped up Lukashenko with cheap oil and gas and demanded loyalty in return. Belarus became Russia's ordinarily faithful vassal, although friction was common and coercion sometimes necessary. When in 2003 Putin proposed the two states merge – Belarus becoming six oblasts of the Russian Federation – Lukashenko flatly refused. Five years later, Putin expected to succeed Lukashenko as president of the Union State Council when he stepped down from the Russian presidency, but Lukashenko objected again, refusing to hand over the office he has held since 2000.

Russian "peacekeeping" in Abkhazia and South Ossetia should be seen in this context too. Something very similar was undertaken in Moldova when, in 1990, Russian-speakers occupying a slither of territory along the country's border with Ukraine rose in armed insurrection against the Moldovan republic. The Soviet and then Russian army aided these "Transnistrian" forces, Russian troops even fighting alongside the so-called "Neo-Cossacks" at times. This conflict lasted two years and cost around 2,000 lives before the Kremlin mediated a peace deal. The arrangement shaved a Transnistrian republic off from Moldova, its final status to be determined later, and established a ceasefire and border policed by Russian "peacekeepers". Russian peacekeeping became a permanent occupation, the leverage Moscow uses to keep Moldova if not within its sphere of influence then, at least, outside the West's.

The *Russkiy mir* may have been on the back foot when Putin assumed the presidency, but Russia's imperial sense of self remained. Putin appealed to it in his letter to the people penned ahead of his 2000 election. There, Putin explained that rebuilding strength at home was his main priority but that this was tied squarely to the ambition of restoring Russia's great power status. Great powers cannot be where "poverty and weakness reign" he wrote. Russia must get its own house in order if it hoped to reclaim its rightful place in the world. As a great power, Russia had a "sphere of vital interests" into which it would seek "expansion in the good sense of the word". No one should fear Russia, Putin explained, but they should "reckon with it".

By happy coincidence, the economic reforms Putin introduced in his first term – which tried to rationalize policy, gather tax and strengthen the state – coincided with a sharp increase in the price of oil and gas. From $32.14 a barrel in January 2001, the price of oil reached $83.95 in January 2007. Consequently, economic growth hit 7.3 per cent in 2003 and stayed at around that level for four years. Incomes rose, quality of life improved. Money poured into the state's coffers. Russia sensibly paid off its foreign debt. By Putin's second term, a consumerist middle class had emerged, violent crime declined, the scourge of alcoholism became less pronounced. National confidence and self-esteem returned. In 2007, Russia was awarded the Winter Olympics (Sochi 2014). In 2010 it was awarded hosting rights for the 2018 FIFA World Cup. Vladimir Putin got the credit for transforming Russia's fortunes. The background story of Putin's first two terms, then, was one of rising Russian confidence. Putin became more confident as he grew into the job. The state became more powerful as the economy grew. Russians looked even more favourably on the man they believed responsible for their good fortune. National pride returned, stoked by the state. In a country that sees itself as an empire with influence and authority extended beyond state borders, a more confident government, a wealthier state, and a more ebullient people made for a more assertive international attitude. The inflow of cash allowed Putin to instigate military renewal. No longer would Russia have to field armies of conscripts and mercenaries. The new Russia would use its financial windfall to professionalize its military, to update its weapons systems, to become a peer to its great rival in the West, the United States. The new Russia was getting the material power it needed to make its imperial space, its sphere of interest, a reality.

For Georgia, this meant that Putin's Kremlin became much less hands-off than Yeltsin's had been in its second term. Yeltsin's laxity had allowed Shevardnadze to engineer more room for manoeuvre. For example, in 1997, Georgia joined the GUAM (Georgia, Ukraine, Azerbaijan, Moldova) grouping to improve cooperation and balance Russia. In 1999, he took advantage of an institutional change to withdraw Georgia from the Collective Security Treaty (something Azerbaijan and Uzbekistan did too). Putin wanted to bring Tbilisi back to the original bargain.

One reason was Chechnya. The Kremlin was concerned that Chechen insurgents and civilians moved relatively freely into Georgian territory (Georgia shares part of its northern border with Chechnya) and wanted permission to conduct military operations there. When the Georgians refused, Russia lifted a ban on the movement of military age men between Abkhazia and Russia initially imposed as part of a raft of measures by the CIS. Although the ban was never vigorously enforced, its lifting conveyed the powerful message that if Georgia did not want to be a part of the Russian institutional space, then it could not expect to receive the privileges of membership. In November 2000, Russia imposed a visa regime on Georgians wanting to travel to Russia but exempted residents of Abkhazia and South Ossetia. Meanwhile, a media campaign in Russia warned that Tbilisi posed a dire threat to Russian-speakers and allies and whipped up anti-Georgian sentiment.

Another set of issues concerned South Ossetia. Elected in 1996, South Ossetia's leader, history professor Lyudvig Chibirov, favoured tentative cooperation with Tbilisi hoping this might facilitate a future peace settlement. Three years later, he and Shevardnadze concluded a roadmap to peace. This threatened the very foundations of Russia's influence over Georgia. To stop it, the Kremlin conspired to orchestrate the election of Eduard Kokoity – a reliably pro-Russian and anti-integrationist former wrestler who espoused integration with Russia. Kokoity was duly elected in December 2001, and in March 2002 the South Ossetian parliament called for the republic's integration into Russia. Three months later, the Russian parliament (Duma) granted Russian citizenship and passports to those inhabitants of South Ossetia, Abkhazia and Adjara who wanted them. Granting citizenship to foreign nationals without their host state's consent is unlawful. It fundamentally recalibrated the political struggles inside Georgia for now the population in the disputed territory included Russian nationals, full members of the *Russkiy mir*.

Escalations continued. In August 2002, Russian aircraft bombed alleged Chechen targets in Pankisi Gorge in Georgia. In early September, Kokoity concluded a military pact with Abkhazia, a move suggested, encouraged and facilitated by Russian officials. By year's end, South Ossetia's defence, police and intelligence ministries were all headed by Russian officials. Putin used the first anniversary of 9/11, to ratchet up pressure by warning that Russia would be within its rights to use force against Georgia since the latter harboured Chechen terrorists. The allusion to George W. Bush's War on Terror could not have been clearer. But instead of buckling, Shevardnadze turned to the US for help balancing Russia. He invited the US to send troops to assist the Georgian military's counterterrorism. With American help, the Pankisi Gorge was cleared of Chechen insurgents. Then, the Americans stayed to support military reform, Georgia's westwards drift becoming clearer. Six weeks after Putin issued his 9/11 threat, Shevardnadze announced his intent to apply for membership of NATO.

From Rose Revolution to Munich

Shevardnadze had rescued Georgia from the brink of collapse in the early 1990s. But his bargain with Yeltsin had not been his only Faustian deal. He had also been forced to make deals with Georgia's oligarchs and militia leaders. As a result, economic and legal reform ceased, corruption prospered, and the economy stagnated. Growth averaged just 3.5 per cent in the five years up to 2003, a figure measured from a very low base after almost a decade of decline. In November 2003, an unpopular government backed by corrupt oligarchs faced parliamentary elections. Given his history, Shevardnadze may have persuaded himself that he alone could hold the country together. Whatever their motives, president and oligarchs conspired to keep the government in power no matter the will of the people. Election-day exit polls suggested that the opposition National Movement led by pro-Western reformist Mikheil Saakashvili had won most votes. But the official count recorded a victory for the government. The electoral fraud was obvious. For example, Adjara, run as a personal fiefdom by the autocrat Aslan Abashidze, posted an implausible 100 per cent vote for the government. Protests erupted in Tbilisi, but Shevardnadze refused to relent prompting running battles between security forces and protestors. An alarmed Putin despatched foreign minister Igor Ivanov to mediate. The Kremlin could see Shevardnadze was losing his grip and feared a popular revolution might push Georgia further from Moscow's orbit. Ivanov proposed that the Georgian leader stay on a few months more and prepare an orderly transition. It was a clarifying moment for the old president, forced to choose between what Moscow dictated and what he knew the Georgian people wanted. He chose the people and resigned. Saakashvili was elected president by a massive majority.

The impact of this "Rose Revolution" reverberated beyond Georgia. The idea that a people might choose to realign their country's international position was an overt challenge to Putin's view of how international relations within "Russia's sphere" should be conducted. Once that idea was out of the box, the Kremlin found it very difficult to suppress. After Georgia's "Rose Revolution" came the "Orange Revolution" in Ukraine (2004–05), the less than successful "Tulip Revolution" in Kyrgyzstan (2005) followed by the 2010 "Melon Revolution" which succeeded in ousting the pro-Russian president Kurmanbek Bakiyev, the suppressed "Denim Revolution" against Lukashenko in Belarus (2006), the pro-European failed "Grape Revolution" in Moldova (2009), and Ukraine's "Revolution of Dignity" (2014) – Euromaidan. Beyond the former Soviet space, Lebanon's 2005 "Cedar Revolution" challenged a key Russian ally, Syria. Behind all these revolutions lay popular demands for democratization, economic reform, an end to kleptocracy and corruption, and in most (but not all) cases closer partnership with the West. Putin saw all this not as manifestations of local democratic struggles against autocracy but as separate parts of one big struggle, between Russia and the West. Each of these uprisings, he found it convenient to believe, was inspired by American conspiracy, funding, orchestration and manipulation aimed at the overthrow of pro-Russian governments and installation of pro-Western regimes. Moscow itself was the ultimate target. Georgia's Rose Revolution was the first to really hit home in Moscow, but as the "Colour Revolutions" spread, so Putin became more convinced that they represented a geopolitical struggle between two rival poles: hegemonic America and its main rival Russia.

The new Georgian government had three main priorities: liberal economic reform, closer ties with the West, and reintegrating the separatist territories. On the first, it made good progress. Economic growth averaged 9.3 per cent between the Rose Revolution and the 2008 war. Corruption declined markedly, Georgia moved from 103rd place in Transparency International's corruption index to 67th. In 2021, it ranked 55th compared to Russia's 136th. On the second, the new government received a cool reception from the Europeans but a much warmer one from US President Bush. To curry the favour with Washington he would need to get into NATO, Saakashvili despatched 2,000 troops to support the US-led coalition in Iraq. It made good progress on the third, initially, too. The Georgian government conducted an effective popular campaign to secure Abashidze's resignation in Adjara. The kleptocrat blew up bridges connecting the territory to the rest of Georgia and turned to Moscow for help. But although he found a friendly ear, the Kremlin was unwilling to offer practical support. Situated on Georgia's southern border with Turkey, Adjara had even less strategic utility to Moscow than South Ossetia, plus Abashidze was a crook almost universally loathed. What is more, Putin had not yet given up on trying to persuade Georgia's new leader to accept the Kremlin's hegemony. In their early exchanges, for instance, Putin offered to help reintegrate South Ossetia into Georgia in return for Saakashvili taking NATO membership off the table. Without Moscow's support, Abashidze could not hold on to power. He fled into exile in Russia and Adjara was reintegrated into Georgia. It was a major early success for Saakashvili.

The Georgian president turned next to South Ossetia. He thought that like Abashidze, Kokoity was an unpopular leader lacking legitimacy and that like Adjara, South Ossetia was not a major Russian strategic interest. But he had seriously misjudged the situation. Saakashvili spoke of reconciliation and moved to suppress Georgian irregular forces. Meanwhile, Georgian authorities increased pressure on Kokoity by clamping down on the separatists' smuggling operations. Rather than undermining Kokoity, however, these measures stoked Ossetian fears. Kokoity's militia fought back and in the summer of 2004 the Georgians were forced to retreat. Fifty Georgian peacekeepers in Tskhinvali were captured and publicly disarmed. Putin meanwhile made it clear that Saakashvili could not expect him to help in South Ossetia as he had in Adjara. South Ossetia's parliament appealed for the Russian Duma to recognize its independence. The Russian government agreed to build new bases, accept South Ossetian soldiers into its military academy, supply additional weapons (including 75 T-72 tanks, artillery systems, and multiple-launch rocket systems), and increase the number of Russian soldiers rotating through the separatist entity. It also suspended the supply of electricity to Georgia. Putin's position hardened further towards the end of 2004 as a result of the Beslan atrocity (just a fortnight after the end of hostilities in South Ossetia) and Ukraine's Orange Revolution. The US and EU urged Georgia to show restraint and not risk military conflict. When Georgia invited the EU to replace Russia as the conflict's mediator, it found the Europeans disinterested. Georgia's bid to reintegrate South Ossetia had utterly failed.

To Putin, South Ossetia and Abkhazia were bargaining chips he could use to influence Saakashvili. If that failed, the two provinces could be attached to Russia by default through a creeping annexation. As well as punishing Tbilisi, creeping annexation would guarantee that

Georgia's bid to join NATO would fail. NATO membership comes with a guarantee (in Article 5 of the North Atlantic Treaty) that the alliance will protect its members from attack. There was therefore no chance the alliance would admit as a new member a state that had an unresolved territorial dispute with Russia. These issues became much more urgent to Putin once Ukraine's "Orange Revolution" prevented Viktor Yanukovych assuming the presidency there. As Putin saw it, what happened in Georgia would most certainly not stay in Georgia. In early 2005, Russia strengthened its influence in Abkhazia by manoeuvring the Kremlin's man, Sergei Bagapsh, into power there. Whatever slim hope there had been of Abkhazia's peaceful reintegration into Georgia was gone.

The Kremlin dramatically increased funding to the two separatist entities, sending more cash annually than they managed to earn themselves, most of it destined for their militaries. By 2006, Russian military assistance to South Ossetia and Abkhazia likely exceeded Georgia's defence budget. Efforts to undermine Georgia abounded. In 2005, Georgian intelligence uncovered a Russian spy ring working to undermine the government. Russia responded by imposing sanctions on the export of Georgian wine, severing transport links, and refusing visas to Georgian nationals. Counterintuitively, it also agreed to close two military bases, one in Batumi in Adjara and one in Akhalkalaki close to Georgia's central southern border with Armenia and Turkey, leased since the end of the Soviet Union, prompting some to hope conciliation remained a possibility. More likely, the purpose was to withdraw Russian forces from isolated bases that could be easily encircled in the event of war and pillaged for Russian arms. No sooner was the agreement to close these bases finalized, than the Russian army was building new bases in the more defensible territory of South Ossetia. The process was completed ahead of schedule in 2007.

By then, Russia's nominees headed the governments of South Ossetia and Abkhazia. Russians staffed the defence and security institutions of both. Both governments were bankrolled by a Kremlin cashed-up by the resources boom. Most residents held Russian passports and travelled freely into Russia. Arms flowed in from Russia and the number of Russian soldiers stationed there was increasing too.

That there was more going on here than the usual rough and tumble of Georgia's difficult relationship with Russia was made clear in Putin's address to the Munich Security Conference in February 2007. Putin's second term in office had reached its final year. Uncertainty as to what would follow reigned, although no one doubted that it would be Putin who decided. His courtiers, principally (but not exclusively) defence minister Sergei Ivanov and Dmitry Medvedev, deputy prime minister and long-term junior aid, jostled for the president's favour. It was in this febrile atmosphere that in October 2006 journalist Anna Politkovskaya – renowned the world over for her remarkable reporting on Russian atrocities in Chechnya – was murdered outside her Moscow home, most likely on the orders of Ramzan Kadyrov. The following month, FSB whistle-blower Alexander Litvinenko was assassinated in London by Andrey Lugovoy and Dmitry Kovtun. Lugovoi, a former KGB/FSB agent, went on to be elected to the Duma, representing Vladimir Zhirinovsky's extreme nationalist "Liberal Democratic Party".

Putin's purpose at Munich was to position himself, and Russia, as a peer and geopolitical rival to American unipolarity: a rival capable of challenging US supremacy in a battle of ideas and material power. The US, he claimed, was bent on subjecting the world to "one single master": itself. American unipolarity was undemocratic, illegitimate. It bred hyper-militarism, plunged the world into "an abyss of permanent conflicts". NATO expansion was a "serious provocation" which aimed to deploy "frontline forces" on Russia's borders. In response, he pledged, Russia would play its historic role as a great power by countering US hegemony. Putin's Russia would champion the pluriverse – a world like that envisaged at Yalta where great powers managed their own spheres of interest and resolved global matters between themselves. To do that, of course, Russia would need to stop its own sphere of influence from melting away.

The five-day war

Russia's creeping annexation of Abkhazia and South Ossetia dates back to Yeltsin but accelerated sharply under Putin who later revealed that military preparations for war began as early as 2006 with the shifting of bases, strengthening of military positions and increasing of aid. Russia's strategy seems to have been to effect annexation without war if possible, but to prepare for war should that be needed. At a minimum, the goal was to prevent Georgia joining NATO but the broader aspiration was to bring Georgia as a whole back into Russia's imperial space. Military preparations and provocations increased during Putin's final year in office. In March 2007, Russian helicopters attacked Georgian government buildings in upper Abkhazia. Georgia's parliament responded by voting unanimously to pursue NATO membership. The US Senate quickly signalled support, which it extended to Ukraine too. In July, Putin unilaterally withdrew Russia from the Treaty on Conventional Force in Europe, citing America's withdrawal from the Anti-Ballistic Missile Treaty as his reason. The most immediate effect, however, was to lift legal restrictions on the number of conventional forces that could be deployed west of the Ural Mountains. Later that month, Russia announced that Abkhazia would be included in infrastructure projects for the 2014 Sochi Olympics. In August, Russian Su-25 jets bombed a Georgian radar station. By November, Russian forces had completed their withdrawal from Batumi and Akhalkalaki and construction of new facilities in Abkhazia and South Ossetia. The following month, Putin anointed Medvedev as his successor, and Medvedev reciprocated by revealing that Putin would head the government as prime minister. Medvedev and Putin called it "the tandem", but no one doubted who would be steering.

As Putin's presidency entered its final five months, two international events seemed to crystallize the confrontation he had spoken about in Munich. In February 2008, Kosovo declared independence from Serbia. Western governments recognized the new state, to Putin another example of dangerous Western unilateralism made more painful by the fact that the country losing territory was Serbia, a Russian ally. Putin had long warned that should the West recognize Kosovo, Russia would recognize Abkhazia and South Ossetia. Now it moved to do just that. In early March, Russia unilaterally withdrew from the CIS embargo on the two regions, allowing it to expand economic, political and military ties openly. Two weeks later, the Duma resolved to begin the process of recognition. This was precisely the sort of unilateralism that Putin himself had denounced in Munich, and it showed that what was at stake was not the question of unilateralism or self-determination but that of spheres of influence. From Putin's perspective, if the West could act as it pleased in Kosovo, then Russia could do the same in Georgia.

The second international event was the April 2008 NATO summit in Bucharest. Among other things, NATO leaders considered the question of membership. Albania and Croatia were to be admitted. Macedonia would have to wait. Georgia and Ukraine's application for a membership action plan (MAP) proved more controversial. The Bush administration threw its weight behind their membership applications, but France and Germany opposed them. German Chancellor Angela Merkel insisted that NATO could not accept prospective members that were parties to territorial disputes. It was a disingenuous claim since NATO already had two members party to a territorial dispute with each other – Greece and Turkey – but it was precisely what Putin expected. The real reason for European wariness was Russia. Forced to choose between their relationship with Russia and the aspirations of Tbilisi and Kyiv, France and Germany chose Russia. The two aspirants were denied MAPs, the crushing blow mollified by a non-binding declaration that, one day, Georgia and Ukraine would join the alliance. NATO's decision to keep Georgia and Ukraine at arm's length was a game-changer. Publicly Putin, his foreign minister Sergei Lavrov, and other senior Russian officials expressed outrage at the alliance's statement, while Saakashvili and Ukraine's President Viktor Yushchenko welcomed it. In private, things were very different. To understand, look not at what was said in Bucharest but at what happened: Georgia and Ukraine asked for a pathway into NATO. NATO said no. Putin understood this perfectly and lost no time communicating his intent, ordering direct relations be established with Sukhumi and Tskhinvali. Saakashvili and Yushchenko left Bucharest bitterly disappointed because they understood what it meant too.

Russian strategy moved from creeping to accelerated de facto annexation, hoping this could be achieved without war but prepared to fight if needed. Between Kosovo and Bucharest, Dmitry Medvedev was duly elected president with a staggering 71 per cent of the vote. Inside Georgia, Russian military preparations gathered pace. Additional Russian troops were deployed to Abkhazia and South Ossetia, Russia's military presence in Abkhazia growing by as much as a quarter. The Black Sea fleet conducted exercises off the Georgian coast. Bagapsh and Kokoity became regular visitors to Moscow. A rail connection between Abkhazia and Russia was hastily repaired – a sure sign of military preparations since Russian armour is moved by rail. Violence increased too.

South Ossetian forces shelled Georgian and mixed-ethnicity villages in late July. They also fired on OSCE observers. At the very least, the South Ossetians were emboldened by Russian support to press their case. It is equally possible given the character of the republic's military leadership, however, that they were instructed to escalate by Moscow. In the first week of August, the Russian military, mainly the 58th Army, performed annual exercises in North Ossetia. "Kavkaz 2008" simulated an intervention to quell violence in Georgia. "Know your enemy" the soldiers were instructed. "The enemy is Georgian". The exercise complete, most of the forces involved remained where they were, just north of the Roki tunnel – the only direct road between North and South Ossetia. On 1 August, five Georgian police officers were killed. Georgian militia responded by killing six South Ossetian officers. Skirmishes erupted across disputed villages; the Kremlin signalled its intent to reinforce its "peacekeeping" mission should the situation deteriorate further.

It is what came next, at the end of the first week of August 2008, that remains a source of intense controversy and was the focus of an independent report commissioned by the EU. Russians and Georgians tell competing narratives that revolve around one central question: had Russian forces in the Roki tunnel already entered Georgia when the Georgian military began its bombardment of Tskhinvali at 11:30pm on 7 August? The Kremlin claims Russia's was a humanitarian intervention launched only after Georgian forces had begun indiscriminately bombing Tskhinvali; the Georgians claimed its military acted in self-defence after Russian forces had deployed into Georgia from the Roki tunnel. It is difficult to piece together precisely what happened on 7 August. Neither leadership entirely engineered the timing. On the Russian side, Medvedev was on vacation on the Black Sea; Putin was in Beijing for the Olympics. But the Russian military *was* thoroughly prepared. The Georgian military was not. A quarter of its principal force was in Iraq.

What seems to have happened was that skirmishes escalated on 6–7 August. Georgian villages, including Nuli, were razed. On the afternoon of the 7 August, Saakashvili, concerned about the escalation, approached the Kremlin for talks. He was refused a direct line to Medvedev or Putin but advised to impose a unilateral ceasefire, which he did. However, South Ossetian forces continued to fire artillery at Georgian positions and villages. At 11.30pm that same day, Saakashvili changed his mind and ordered the assault on Tskhinvali. At this point, there had not been a major deployment of Russian forces through the Roki tunnel, but a creeping deployment that would nonetheless achieve the goal of de facto annexation seemed likely. The US counselled restraint but advised it was unlikely that Russia would intervene unilaterally. Believing de facto annexation all but inevitable unless he acted, Saakashvili gambled on a lightning assault to create new facts on the ground the Kremlin could not ignore. His gamble was that Georgian forces could take Tskhinvali before the Russians could react. But both he, and the US, had underestimated Putin's intent and the military capability Russia had amassed on the northern side of the Roki tunnel.

When Georgian artillery opened-up on Tskhinvali at 11.30pm they expected a short, sharp, victory over the South Ossetians and entered the town that morning. But before 6.00am, the Russian army was exiting the Roki tunnel into Georgia. Soon after, Russian aircraft targeted advancing Georgian forces – apparently without having been specifically ordered to do so by either Medvedev or Putin, suggesting plans had already been approved.

Russian artillery and airpower pushed the Georgians out of Tskhinvali within a couple of hours. It was Putin from Beijing, not Medvedev – now back in Moscow – calling the shots. Putin issued his public statement before the president issued his, and it was the prime minister not the president who rushed to join Russian forces at their staging ground in North Ossetia on day two of the war. But things were not going entirely to plan. A Georgian counterattack on Tskhinvali prolonged the fight there into the late evening. The following day, Russia's war widened dramatically. Aircraft attacked Georgian military bases and airfields across the country, whilst armour, artillery and troops moved into Abkhazia on the recently rebuilt railway line. Stray shells hit an apartment block in the central Georgian city of Gori.

On the third day, 4,000 Russian paratroopers landed in southern Abkhazia to open a second front. By now, Russian and South Ossetian forces controlled all of South Ossetia. Overpowered Georgian forces were in disorderly retreat, accompanied by 40,000 Georgian civilians. The patchwork quilt of Georgian and Ossetian villages that had characterized the frozen conflict for nearly two decades was being transformed, most likely permanently, by war.

On day four, Russian forces advanced beyond South Ossetia towards Gori, just 86 km from Tbilisi. Georgian forces abandoned the town and retreated to make their stand outside the capital. On the fifth day, the Russians advanced beyond Gori (surrounding but not taking the city) to within 30 km of the capital. They also moved southwards out of Abkhazia, whilst marine and naval operations took towns and ports along Georgia's Black Sea coast, notably Georgia's largest port, Poti. Perhaps surprised by just how decisive and swift the victory, Putin toyed with widening his objectives to include the removal of Saakashvili. Lavrov confided as much to US Secretary of State Condoleezza Rice, who stridently warned against it. The US stated it would not intervene directly, even though Georgia had proven itself a loyal and committed ally. Nonetheless, it airlifted the 2,000 troops Georgia had deployed in Iraq back to defend the capital.

If Russia wanted to remove Saakashvili, it would have to storm Tbilisi as it had stormed Grozny. To this point, the war had cost relatively little. The Russians lost fewer than 70 soldiers killed and around 250 wounded. The Georgians lost more, 170 killed and close to 1,000 wounded but still relatively light compared to what had gone before and what would come later. The invasion had also attracted relatively light international condemnation thanks largely to confusion over who had started it. Avoiding cities had also kept civilian casualties low: Russia claimed 160 civilians were killed in South Ossetia by the Georgians; Georgia claims 220 civilians were killed by the Russians. An assault on Tbilisi to change the regime would have changed that. It might have also exposed problems inside the Russian military. So far, combat performance had appeared so impressive in part because Georgian resistance was so weak. The Russians still faced chronic problems in command and control, coordinating the use of air power, the quality of its equipment, and the professionalism of its troops that might all be exposed by a protracted siege. Putin seems to have thought the better of it and decided to stick to his original intent.

French President Nicolas Sarkozy shuttled between Moscow and Tbilisi to negotiate a deal that gave Putin everything he wanted: the complete and permanent withdrawal of Georgian troops, police and officials from Abkhazia and South Ossetia, and the permanent stationing of Russian forces there. Russian forces withdrew gradually from Georgia proper and established a permanent presence of around 10,000 military and FSB personnel in the two republics. In late September, Medvedev signed off on the Duma's decision to grant diplomatic recognition to

Abkhazia and South Ossetia. The move was widely criticized in the West and created some discomfort in Russia's relationship with China, but there were few penalties or sanctions. European leaders wanted to avoid singling out the Russians for blame. The EU's independent report, crafted by a team led by Swiss diplomat Heidi Tagliavini, found fault on both sides, but ultimately blamed Georgia for starting the war. Putin took it as vindication of his strategy but also as a sign that a divided and chastened West would do little to challenge the new Russia inside its own sphere of influence.

Putinism triumphant

The five-day war appeared a glorious victory for Putinism. Russia achieved all its objectives at minimal costs thanks to a rapid and decisive display of force. Recognized as independent states, Abkhazia and South Ossetia were tethered to the Russian Federation. Above all, however, Georgia's aspiration to escape Russian influence and forge new ties with the West was dealt a devastating blow. Russia emerged with its self-image as a great power burnished by military success. It was a success much clearer and more decisive than Putin's Chechnya war, one that seemingly laid to rest the ghosts of past military failures. It spoke loudly of Russia's return as a great power and was a message others were meant to hear; a warning to anyone else who might think of wriggling away from Russian power.

Putinism crystallized as more than just a policy or ideology, but as a vision of Russia, that recalled the Russian empire and that now stood vindicated, a vital component of Russia's new identity. Russia's enemies had not just been defeated, they had been humiliated. Not since Berlin 1945 had that been quite so obviously achieved. Inside Russia, victory was greeted with a public hysteria born of national pride and confidence, a sharp counterpoint to the despair of the 1990s. The architect of Russian revival, Vladimir Putin, began to get a clearer sense of his own purpose, his historical mission, a mission alluded to in Munich but still not yet fully worked out. In sharp contrast, Western feebleness was exposed for all to see. The US and EU big on rhetorical bluster but fatally short of political unity, will and power. Saakashvili, a figure likened to Hitler by Russian state media, had promised his country much but had been outwitted and outgunned. Apparent Western weakness, its willingness to abandon Georgia in Bucharest, blame Georgia for the war, and underwrite a settlement that awarded Russia everything it wanted, fuelled Putin's confidence. Putin saw war in Georgia as one part of the broader struggle he had spoken about at Munich – a struggle between a hegemonic West and Russia. Georgia wasn't the only democracy that wanted to extricate itself from the *Russkiy mir*. A much bigger opponent resided in what Putin thought to be the very heart of the Russian world. In Ukraine.

Further reading

Ronald Asmus, *A Little War that Shook the World: Georgia, Russia, and the Future of the West*. London: Palgrave Macmillan, 2010.

Agnia Grigas, *Beyond Crimea: The New Russian Empire*. New Haven, CT: Yale University Press, 2016.

Thomas de Waal, *The Caucuses: An Introduction*, second edition. Oxford: Oxford University Press, 2019.

Anna Mouritzen and Anders Wivel, *Explaining Foreign Policy: International Diplomacy and the Russo-Georgian War*. Boulder, CO: Lynne Rienner, 2012.

Anastasia Shesterinina, *Mobilizing in Uncertainty: Collective Identities and War in Abkhazia*. Ithaca, NY: Cornell University Press, 2021.

4
Ukraine I

Vladimir Putin insists Ukraine is not a real country. That its people are really Russians. That its state is not fully sovereign. That its boundaries are historical accidents. Russia's president has said this often. His reasoning evokes such an unadulterated imperialism of earlier centuries it is hard to see how a modern leader could be more overtly imperialist. But it is not just Putin. From what we can tell, many Russians think the same way. When, in the 1990s, Boris Yeltsin's reform-minded foreign minister, Andrei Kozyrev, talked of reintegrating Russia with former Soviet republics, it was Ukraine and Belarus he had in mind. For Kozyrev as for Putin, it was axiomatic that Russia's identity as a great power hinged on a hierarchical relationship with its post-Soviet neighbours, typically referred to as the "near abroad" by Russians to denote that these states were not properly "foreign" or sovereign. Russian control of Ukraine is an article of faith for many Russians, rooted in the idea that Russians and Ukrainians are one people. Held by nationalists, communists, and many liberals, this attitude towards Ukraine plainly exhibits Russia's imperial sense of self.

In the ninth century, adventurous Vikings calling themselves the "Rus" established riverine trade routes between Scandinavia and Byzantium. One of their greatest was on the mighty Dnieper River. A staging post, Kyiv evolved into the centre of a prosperous warrior kingdom. Kyivan Rus expanded over the next couple of centuries, at its peak extending from the Black Sea to the Baltic. In the late tenth century, its ruler, Prince Volodymyr (Vladimir in Russian) converted to Eastern Orthodoxy. Under the leadership of Prince Yaroslav "the Wise", Kyiv erected its own Santa Sofia cathedral in the eleventh century. Yaroslav's children were less than wise, however. They squabbled and fought internecine wars that divided the territory and weakened themselves.

The end came at the hands of the Mongols in the thirteenth century and it was here that the Rus became two or perhaps three peoples. Russian nationalists claim that when Kyiv fell, the Rus fled north to Muscovy and established there the successor to the Kyivan Rus. Ukrainian nationalists tell a different story, that the Rus remained in Kyiv and founded Ukraine. The truth is lost to history, but most likely some fled, and some stayed. More important was the fact that for the next 400 years, Kyivans and Muscovites lived in different political orbits. Mongol occupation of Kyiv endured only a few years after which Kyiv and the lands to the west came into the orbit of pagan, then Orthodox, Lithuanians and Catholic Poles, a part of the Polish-Lithuanian Commonwealth that endured until the eighteenth century. To the north, Muscovy remained a Mongol vassal for more than 200 years. The duchy had already freed itself from waning Mongol power when the Romanovs assumed power in the early seventeenth century and turned to their own imperial project. A few decades later, Ukrainian nobleman Bohdan Khelmnytsky led an uprising against the Poles and turned to the Tsar of Muscovy for help. The proposed alliance appealed to a Moscow eager to win the land and wealth it believed necessary to protect itself from rapacious Poles to the west and barbarous Mongols to the east. In trying to save Ukraine from the Poles, Khelmnystky handed it to Moscow. Ukraine became "Little Russia". In the eighteenth century, Catherine the Great added Crimea to the Russian empire, forcibly seizing it from the Turkic Tatars who had governed for centuries under Ottoman suzerainty. Catherine also brought the wild steppe east of the Dnieper into the Russian empire, part of what imperialists called *Novorossiya* – "new Russia" (part of which is now known as Donbas). The empire made a concerted effort to Russify its new lands, and it almost succeeded. Thus is Ukraine embedded in the mythos of Russian nationalism; a nationalism resting on an imperial sense of self.

At the start of the twentieth century the lands of modern-day Ukraine were split between Russian and Habsburg empires. The 1917 revolution put paid to the empire of the Tsars, but a short-lived Ukrainian republic was soon overrun by the empire of the Bolsheviks. The first few decades of Bolshevik rule were catastrophic. To the Bolsheviks, the peoples of Ukraine were doubly suspicious: their national allegiance was ambiguous and rural Ukraine's way of life anathema to Marxist-Leninism. Socially conservative small-property holders, Ukrainian kulaks were everything the Bolsheviks despised. "Dekulakization" involved mass deportations and mass executions, the aim nothing short of liquidation. Millions were deported to Central Asia and Siberia where a staggering proportion died. At least 500,000 were shot. The state-made famine this provoked, the Holodomor, killed approximately five million Ukrainians. While Ukrainians starved, wheat continued to be exported not just to Moscow but abroad. Ukraine then bore the brunt of Hitler's Nazi invasion in 1941. Caught between two of history's bloodiest tyrants, more than six million Ukrainians, including more than one million Jews, perished. The whole of Ukraine fell under Russian Soviet control for the first time and things settled down after the war. In 1954, Khrushchev transferred jurisdiction over Crimea from Russia to the Soviet Republic of Ukraine, most likely to win Ukrainian support for his leadership bid. To Russians like Putin this was a historic betrayal since to them Crimea is quintessentially Russian. The fact Crimea was annexed by Russia only in the eighteenth century, becoming majority Russian only in 1944 when Stalin forcibly deported the Tatars didn't matter.

Orange Revolution

Ukraine played an important part in the demise of the Soviet Union. After the August 1991 coup, Ukrainians – including majorities in Crimea and the Donbas – voted overwhelmingly for independence, effectively blocking Yeltsin's attempt to supplant the Soviet state with a Russian-led successor. Ukraine's last communist leader, Leonid Kravchuk, then conspired with Yeltsin to end the union. Kravchuk understood that the basis of political power had shifted, that it was the nation that mattered now, and the nation – Ukrainian and Russian-speakers alike – overwhelmingly preferred independence to Russian suzerainty. He agreed to bring Ukraine into the CIS but persistently opposed Yeltsin's attempts to make it more than a loose association and asserted Ukrainian sovereignty at every opportunity. The question of whether the relationship between Russia and Ukraine should be a horizontal one between sovereign peers or a vertical one between hegemon and subordinate was thus a bone of contention from the outset.

There were two other bones of contention as well: Ukraine's inheritance of that part of the Soviet Union's nuclear arsenal stationed on its territory and ownership of the Soviet Black Sea fleet and its base at Sevastopol. The nuclear issue was resolved by a 1994 agreement in which Ukraine agreed to trade its nuclear weapons for a Russian and American assurance of its territorial integrity. The naval issue was managed by the transfer of the bulk of the fleet to Russia and the granting of a lease to use the base at Sevastopol in return for further security guarantees and favourable pricing of Russian gas. Ultimately, the two issues were managed because neither side had the will or capacity to pursue confrontation and both had bigger issues to deal with.

At home, Kravchuk proved a reluctant reformer. Ukraine's first president barely loosened the state's grip on the economy. If Russia in the 1990s was a posterchild for the shortcomings of economic shock therapy, Ukraine proved that going slow was even worse. Price liberalization and limited legislative reform and privatization led to massive inflation. By 2000, Ukraine's GDP per capita was well below half what it had been ten years earlier. Corruption was even worse than in Russia, Ukrainian industry and state capacity even more decrepit. One thing Ukraine did have, however, was political pluralism. Whereas Russian oligarchs mobilized behind the Yeltsin "family", their Ukrainian counterparts mobilized into different competing blocs. Evidence of that came in 1994 when Kravchuk was narrowly defeated by Leonid Kuchma, whose platform of strengthening economic ties with Russia won him the support of economic elites and Ukraine's industrial heartland in the Donbas. In power, however, Kuchma hewed a line not dissimilar to his predecessor. Privatization continued to be half-hearted and highly corrupt. In foreign policy, although Russian-leaning, Kuchma proved no more willing than Kravchuk to yield sovereignty to the Kremlin. Instead, he forged ties with both Russia and the West, keeping Ukraine equidistant from both but over time tilting towards the West whilst holding off Russian entreaties for Ukraine to join the CIS customs union. Kuchma tried to temper Russian hegemony by coordinating with Georgia, Azerbaijan and Moldova through the GUAM initiative. Ukraine also joined NATO's partnership for peace and deepened its relationship with the EU. Kuchma sought security and market access from the West but since he had no interest in yielding to European values such as the rule of law, anti-corruption, and human rights cooperation remained limited. As his government became more authoritarian, Ukraine's relationship with Europe cooled.

Russian primacy in its "sphere of privileged interest" was the cornerstone of Putin's foreign policy from the start. This imperial vision became more coherent and more important with time. Putin offered to crown Kuchma head of the CIS Council in return for Ukrainian membership of the customs union. Putin saw Ukraine as a core element of his Eurasian Economic Community, and later Eurasian Union, schemes. Neither project made much sense if Ukraine was not involved. Meanwhile, Russia exerted influence by supplying energy at less than market price through a corrupt system of intermediaries that made some oligarchs on both sides extremely rich. As Ukraine's economy worsened and Russia-dependency grew, Kuchma became more willing to accept Russian hegemony but found that pathway repeatedly blocked by Ukraine's Rada (parliament), which still bristled at the implied loss of sovereignty.

This contest over Ukraine's political identity came to a head in the 2004 presidential election – an election conducted in a context of deteriorating relations between Putin and the West following the Rose Revolution, US invasion of Iraq (2003) and the further enlargement of NATO to include Estonia, Latvia and Lithuania (2004). Content with the pro-Russian direction of Ukrainian policy, Putin supported the status-quo candidate. Term limits meant that could not be Kuchma, so the Kremlin backed Kuchma's and Ukraine's eastern oligarchs' choice for president: Viktor Yanukovych, a graceless Donetsk-based heavy with a criminal record for robbery and assault. Putin regarded Yanukovych with contempt, but preferred him mightily to the alternative, Victor Yushchenko, an urbane banker and liberal reformist. The former would keep faith with Moscow; the latter was unabashedly Western in orientation. Yushchenko's platform spoke of pursuing EU and NATO membership. The stakes for Putin were clear, especially coming so soon after the Rose Revolution.

Putin invested heavily in Yanukovych's election, sending his top "political technologists" – led by chief ideologist of Putinism, Vladislav Surkov – and tens of millions of dollars, as well as campaigning directly on the Ukrainian's behalf. Putin promised Ukrainians a preferential visa regime and cut-price energy if they voted for Yanukovych. But Putin and Surkov badly misjudged the situation. First, they underestimated the degree of antipathy the old government aroused. Ukrainians were much more uncomfortable than typical Russians about endemic corruption and their government's slide towards authoritarianism. The tipping point for Kuchma was the assassination of journalist Georgiv Gongadze. Second, they misjudged the degree of political consensus on the defence of Ukrainian sovereignty. Ostentatious displays of Putin's ardour might please Russians, but it worried many Ukrainians. Third, they misjudged the strength of Ukraine's political pluralism. However bad their governments, Ukrainians had become used to free elections and were not prepared to relinquish them. Pluralism was helped by the fact Ukraine's oligarchs were disunited. Although most were inclined towards Russian-style economics, they also valued their independence and feared Putin might try to bring them to heal as he had Russia's oligarchs. The other thing that united the people against Putin's candidate was the poisoning of Yushchenko, probably by Ukraine's own security services. Yushchenko survived the dioxin and, although in obvious pain, returned to the campaign more popular than ever, his scarred face a symbol of the old regime's turpitude.

Fearing defeat in a fair election, Putin's political technologists schooled Yanukovych on how to steal the vote. Ballots were stuffed and absentee ballots fabricated, producing an official result wildly at odds with exit polls. The vote steal was comedically obvious – in Donbas, turnout was reported as a ludicrous 96 per cent in Donetsk and 90 per cent in Luhansk. Kyivans refused to accept the fraud and poured onto the city's Maidan Square to protest. Ukraine's oligarchs and the EU swung behind them. Unwilling to risk his own future for a candidate he had no regard for, Kuchma chose not to order a crackdown. Instead, he asked the EU and Russia to mediate. The parties agreed to rerun the election under tight international monitoring. When they did, Yushchenko won comfortably.

Ukraine's Orange Revolution was an unwelcome shock to Putin. Russians were stunned to discover that Ukrainians might prefer Brussels to Moscow. Putin had assumed that what worked in Russian electioneering – flattery, bribery, coercion and theft – would also work in Ukraine but he was wrong. Ukraine was not Russia and its political culture had developed very differently. Unable to admit his mistakes, Putin reacted as he had done to Georgia's turn by blaming the West. He concluded that the Orange Revolution was not an expression of Ukrainian popular will but an American conceived conspiracy to weaken Russia. To Putin, colour revolutions were a new form of warfare, a way in which the global hegemon used its power subversively to advance its interests at Russian expense. Surkov agreed and presented colour revolutions as a grave threat to Russian sovereignty. What Robert Horvath called "preventive counter-revolution" became a core component of Putinism at home and abroad. At home, Vladislav Surkov established *Nashi*, a Putinist youth group staffed with minor thugs to counter anti-government protests. Legislation curbed foreign funding for civil society groups. Abroad, Russia would have to do more to support its allies and weaken its opponents.

Euromaidan

Coming a decade after the Orange Revolution, the Euromaidan crisis of 2013–14, Ukraine's "Revolution of Dignity" was a two-part play. The first part was a Ukrainian political drama about the country's identity and future orientation. In this, the EU and Russia were supporting actors, albeit important ones. The second part was a war of imperialism-fuelled international aggression. This part was all about Putinism – and its limits in the face of resolute Ukrainian resistance.

Coming so soon after Georgia's Rose Revolution, Ukraine's Orange Revolution had confirmed Putin's belief that the US and its Western allies were waging clandestine war on the Kremlin, one waged not primarily by tanks and planes but by subterfuge and popular protests. Assuming everyone else saw the world as he did, Putin judged that the West was trying to expand its hegemonic space by toppling Russia-friendly governments and enlarging NATO and the EU. Whereas Russian rhetoric tended to focus on the former, the threat posed by the latter was seen as no less insidious. Indeed, it was principally the EU, not NATO, that endangered Putin's Eurasian ambitions. This zero-sum logic made little sense to Western ears and bore no relationship to political reality. The EU and NATO were cautious enlargers who erected tall barriers for prospective members to climb. For both, enlargement was a primarily demand-driven exercise, the demand coming from democratic peoples desirous of the prosperity and rule of law associated with the EU, the security associated with NATO, and the democracy associated with both. That Russia's leadership saw things different speaks volumes about its imperial mindset, which assumed only great powers had agency and authority, that weaker states were vassals to the powerful, and civil societies mere pawns to be manipulated.

In the decade following the Orange Revolution, three trends pulled Russia and Ukraine towards a moment of crisis where Ukrainians would have to choose between Moscow and Brussels. The first was the worsening relationship between Russia and the West. Conflict over the West's recognition of Kosovo, the extension of America's anti-ballistic missile programme into Europe, and NATO enlargement have already been mentioned. Russia's war on Georgia, the NATO-led and UN authorized intervention in Libya in 2011, and sharp disagreements over Syria, polarized relations further. Periodically, Western leaders tried reaching out. The EU offered to bring Russia into partnership arrangements; Germany collaborated on the Nord Stream gas pipelines; the Obama administration offered to "reset" bilateral relations and scale back and slow down the missiles programme. These gestures made little difference to the overall trajectory of relations between Russia and the West. The principal issue dividing them was the colour revolution and the political orientation of governments in what the Kremlin judged to be its sphere of influence. After the Orange Revolution, Putin saw plenty of evidence to support his thesis about Western agitation, including two failed revolutions in Kyrgyzstan (2005 and 2010), a failed uprising against Lukashenko (2006), and a successful uprising in Moldova (2009) that upturned the fraudulent election of pro-Russian communists. The rapid demise of longstanding authoritarians in Egypt, Tunisia and Libya at the start of the Arab Spring in 2011 and the revolution in Syria was simultaneously a shock to the Kremlin and further confirmation of Western perfidiousness.

The Kremlin's uncertain reaction to events in Libya in 2011 shows it was Putin himself driving this agenda. Medvedev was president when the Libya crisis hit, and it was Medvedev who decided that Russia should allow the UN to authorize intervention to protect civilians from Muammar Gaddafi's brutality. Medvedev judged that since Russia had few direct interests and since intervention enjoyed Arab League support, it should not stand in the way. Putin thought differently and openly criticized the president. The issue was that for Medvedev, Libya was a discrete problem of limited importance, whereas for Putin it was intrinsically connected to a broader political struggle between Russia and the West.

By 2009, the EU had developed an Eastern Neighbourhood Policy (ENP) to bring non-members in the east, including Ukraine, Georgia, Belarus, Moldova, Armenia and Azerbaijan, into closer partnership with it whilst setting the question of membership to one side. Putin was having none of it. Where the EU understood ENP as an alternative to membership in the medium term, Putin saw it as a rival to his Eurasian schemes. He made it plain that states in the Russian world should not join the ENP and strived to offer an alternative. That came in the form of a customs union with Belarus and Kazakhstan announced in November 2011 and plans to establish a Eurasian Economic Union, later the Eurasian Union. This effectively ruled out the possibility that Europe might develop overlapping and interlocking systems of economic and political partnership. Instead, states would face a zero-sum choice between the EU and whatever options Moscow was offering.

The second trend was Russia's own authoritarian slide. Putin's decision to return to the presidency ruptured Putinism-by-consent. It was not that Putin was unpopular. He was by far Russia's most popular politician. The problem was that his decision to return to the presidency and claim it had been planned all along unmasked the phoniness of Russian democracy. It demonstrated that it was Putin, and only Putin, that mattered. That the election rallies, the voting, the television chatter, had all been a charade. Polls showed support for Putin ebbing away. Trust in the prime minister fell below 50 per cent; his personal approval – stratospheric in the heady days of the Georgian war – hovered around 60 per cent. Half of Russians polled in late 2011 expressed disappointment with the government which Putin, as prime minister, led; only a quarter believed it had navigated Russia's challenges successfully. Putin's political technologists got to work. They smeared, harassed and imprisoned opponents. They prevented real challengers like Alexei Navalny from standing. Businessman Mikhail Prokhorov ran as a liberal candidate for the Right Cause party but when he realized the party was a front organized by Surkov, and not a genuine Mcdvedev-leaning alternative to Putin, he resigned and ran as an independent. When Putin's United Russia party secured less than half the votes in the December 2011 Duma elections, Putin and Surkov decided to simply steal the presidential vote the following March. The theft was obvious. Massive ballot stuffing led to ridiculously inflated turnout figures in some parts of the country but delivered Putin the victory he wanted.

The winter of 2011–12 was a winter of discontent. People took to the streets in huge numbers right across Russia. The state responded. Opposition leaders were imprisoned, protestors picked up by the hundred, laws were tightened. Causing offence to holders of public office was made illegal. Putinists are easily offended. The protests subsided but there was no mistaking the rupture to Putinism-by-consent. Like never before, Putinism needed the jackboot of authoritarianism to keep political opponents down. The social contract between president and people needed revision. Alongside more overt authoritarianism, the new deal had two main strands. The first was populism. The last time Muscovites had taken to the streets in large numbers was in 2008 when Medvedev attempted to wind back spending on pensions. Now, Putin bought consent by directing a massive increase in pensions and social support. Over the coming decade, as much as half Russia's financial reserves, amassed through the sale of natural resources at high prices, was spent on welfare payments to shore up support for the government. The second strand was nationalism. Putin's second presidency needed a purpose. A reason why *he* should be president. Playing the anti-Yeltsin card was no longer sufficient. Nationalism provided the key. With even greater vigour than before, Putin positioned Russia as an exceptional great power with distinct values and a "sphere of privileged interest" – an imperial space, and himself as the man destined by history to lead. Russia's hosting of the 2014 Sochi Olympics and 2018 football world cup were powerful signs the world recognized Russia's greatness. These events were lavished with presidential care and an unseemly amount of money. The Sochi games cost at least ten times more than the Vancouver Winter Olympics and 30 per cent more than the much bigger summer Olympics held in London. Third-term Putin became more authoritarian, more populist, and more nationalist. It was an incendiary mix.

The third trend was in Ukraine itself. As in 2004, the Euromaidan crisis of 2013–14 was a primarily Ukrainian affair. That it erupted into so much more owed everything to Russia's growing authoritarian nationalism. The Orange Revolution had inaugurated a government that pursued closer partnership with Europe. Yushchenko and his prime minister Yulia Tymoshenko espoused economic and political liberalization and sought membership of NATO and the EU. But their coalition, and with it the government's reformist zeal, quickly deteriorated into such a bitter struggle that the president fired his erstwhile ally and appointed his former opponent Yanukovych. Yushchenko's legislative agenda stalled, his problems compounded by opposition from Ukraine's oligarchs who may not have wanted to be dominated by Putin were equally resistant to regulation by Kyiv. Many were now in positions of authority and used those positions to protect their interests and block reforms. The 2008 global financial crisis hit Ukraine's unreformed economy hard. GDP declined 15 per cent in a single year. All this added to Western reticence towards deeper partnership so little progress was made towards membership. Ukraine suffered the same fate as Georgia at NATO's 2008 Bucharest Summit. Yushchenko's political stocks tanked.

Putin was far from inactive meanwhile. The Kremlin applied what pressure it could to undermine Yushchenko's government. In 2005, it more than doubled the price of gas. Yushchenko agreed to pay the new price but asked for a gradual transition. Moscow refused and in the middle of a cold winter, shut off supplies. Kyiv responded by siphoning off gas conveyed to EU countries via pipelines traversing Ukraine. This took a toll on Gazprom's profits and annoyed some of its best customers, so a deal was struck to get Ukraine's gas flowing again. But the Kremlin wanted to remove this block on its influence, so Russia invested heavily in the Nord Stream project, a pipeline under the Baltic Sea that could supply gas directly to Western Europe and cut out the Ukrainian intermediary. By 2009, Ukraine had amassed a Gazprom debt of more than $2 billion which it was in no position to repay, its economy still in freefall after the financial crisis. Putin ordered that supplies be stopped and the flow to EU countries slowed to inhibit siphoning. Once again, the crisis cost Gazprom – $1 billion by some estimates – but demonstrated just how beholden to Putin Ukraine and the EU had become.

The promise of the Orange Revolution lay in tatters. In 2010, Yushchenko and Tymoshenko ran against each other for the presidency. Both were beaten by Yanukovych. The country's economic plight was probably sufficient explanation for the result, but government ineptitude didn't help. Yanukovych promised healthier relations with Russia and with it a better deal on gas prices. Putin was only too happy to oblige, in return for an extended lease for the Sevastopol naval base, which Yanukovych was happy to grant. From Putin's perspective, Yanukovych was an ideal president, another Lukashenko, and the perfect antidote to Western hegemony. Yanukovych made Russian an official language. He also followed Putin's lead to ensure he could hold on to power. Legislative authority and executive appointments were concentrated in the presidency, the judiciary tamed to the president's will. Yanukovych used these powers to bring opponents and oligarchs to heal. Tymoshenko was jailed for financial impropriety whilst the president corruptly amassed a vast personal fortune. On a 350-acre site at Mezhyhirya, he built a personal pleasure palace replete with private zoo, golf course and galleon. Transparency International rated Ukraine as even more corrupt than Russia.

But on one issue, the Ukrainian president was recalcitrant. Like Kravchuk and Kuchma before, Yanukovych refused to tie Ukraine into Russia's customs union and Eurasian Union projects, thinking he could get a better deal by playing Russia and the EU off against each other. He wanted cheap gas from Russia and investment and market access from the EU. The problem was that Putin didn't want a pluralist horizontal partnership but a vertical association that united the *Russkiy mir* under Russian leadership. Kazakhstan and Belarus joined Russia's single market, amid its winter of discontent. That made Ukraine stand out all the more. But the single market was just a step. Putin wanted a Eurasian Union to institutionalize Russia's imperial space and establish itself as a rival to the European Union.

It was in this context that the EU launched its ENP. Seemingly oblivious to Moscow's zero-sum mentality, European leaders saw no reason why states like Ukraine and Armenia might not enjoy partnership with Europe *and* Russia. After all, they intended the ENP as an alternative to EU membership not a pathway towards it. After years of ambivalence, under the ENP the European Union committed to negotiating an association agreement with Ukraine that would include a "deep and comprehensive" free trade agreement. Yet the two unions – European and Eurasian – were not very alike. They differed on scale (the EU being the world's third largest market, the Eurasian Union having a market a little larger than Italy's), values (the EU insistent on democracy and the rule of law, the Eurasian Union not) and power structure (no single state dominated the EU, Russia clearly dominated the Eurasian Union).

The Europeans and Russians understood the stakes very differently. EU negotiators made half-hearted attempts to bring Russia into their negotiations with Ukraine but that didn't get far. They saw themselves as representing a civilian not military power pursing an entirely consensual partnership. To them, Ukraine's choice need not be zero-sum. Ukraine could accept Europe's terms or reject them; and it could pursue closer ties with Russia irrespective (notwithstanding the fact that it would be a complex technical challenge to reconcile a free trade arrangement with the EU and a customs union with Russia). National leaders in France, Germany, Italy and the UK, meanwhile, were ambivalent about the whole thing – an agreement would be welcomed if concluded but not much lamented if not. Moreover, agreement would be reached only if Ukraine accepted European values. Tymoshenko's imprisonment was a particular sticking point in that regard. The worldview of Western Europeans was very different to Putin's and they struggled to comprehend it. Russia's position was more straightforward: Yanukovych should join the customs union and reject the EU. If it did, Moscow would supply cheap gas and not care who Yanukovych imprisoned. But if it didn't, Kyiv could expect punishment.

Things got more serious in the summer of 2013 when Ukraine and the EU reached a compromise on Tymoshenko that would see her released but exiled to Germany on the pretext of receiving medical treatment. The way was open for a deal, so Putin ratcheted up the pressure. In August, Russia imposed restrictions on the importing of Ukrainian goods to signal the price Ukraine would pay for choosing the EU. That was immense since Russia accounted for a quarter of Ukraine's and when combined with the threat of hiked gas prices it endangered Ukraine's economy with near-immediate bankruptcy. Unsurprisingly, this hardened the battlelines in Ukraine. To many Ukrainians, Putin's measures proved that economic union with Russia meant surrendering sovereignty to Moscow. Aware it should offer carrots as well as sticks, the Kremlin proposed a financial aid package that far outstripped anything the EU was offering. Meanwhile, Russia's chief negotiator Sergei Glazyev – a pronounced nationalist who once described Ukraine's aspiration to join the EU a "sick self-delusion" – issued dire warnings about the consequences that would befall Ukraine if it turned its back on Russia. Debts would be recalled, gas charged at world prices or cut off altogether, Ukraine forced to default and freeze, commitments to respect Ukrainian sovereignty torn up. Backed into a corner, Yanukovych walked away from the EU a week before the association agreement was due to be signed and concluded a deal with Russia instead.

The president's about-turn provoked an initially modest public response. Just a few thousand protestors turned out on Maidan Square, not enough to change Yanukovych's course let alone topple him from power. Not that this was even their goal. Protestors wanted a change of policy not government. Their character were transformed, however, by the president's own decision to use force, something Kuchma had refused to do in 2004 and which Putin now urged Yanukovych to do to prevent a second Orange Revolution. In the early morning of 30 November, the Interior Ministry's elite Berkut forces, many of them recruited from Crimea and the Donbas, used batons and stun grenades to disperse protestors encamped on Maidan Square. The violence sparked popular anger and the protests swelled. Demonstrators reoccupied the Maidan, establishing a tent city. Their demands widened. Now it *was* about the government. But still the protests were not sufficiently large to threaten it.

In mid-December, Yanukovych and Putin agreed that in return for walking away from the EU and extending the Black Sea fleet's lease in Crimea, Russia would bail out the Ukrainian economy to the tune of $15 billion and sell it gas at half the world market price. The Russians again called on Yanukovych to suppress the protests, Glazyev claiming that since the US was arming protestors, Russia was duty bound to protect the government. In mid-January, the Yanukovych government forced a raft of new laws through the Rada which targeted protestors and restricted political freedoms. This backfired too. Dubbed the "dictatorship laws" and combined with the cosying up to Putin, the rhetoric coming from Moscow, and police violence, they signalled a further raising of the political stakes. What began as a protest against a single policy and had widened into a movement against the government and had now widened still further into a struggle over Ukraine's very identity. Was Ukraine to be a European or Eurasian society? Sovereign or dependent? Democratic or authoritarian?

In mid-February, the protests showing no sign of abating, the security forces again used force, this time rubber bullets, stun grenades and tear gas to stop a group of 20,000 marching on parliament. Some protestors, some but by no means all mobilized by far-right groups, retaliated with stones and Molotov cocktails. When protestors attacked police lines on 20 February, Berkut troops fired back with live ammunition. More than 70 protestors and a dozen security officers were killed. In total, according to the UN, 108 protestors and 13 security officers were killed in the three months between late November and late February. The violence crushed what little remained of the government's legitimacy. Forced into retreat, Yanukovych accepted a deal brokered by the foreign ministers of France, Germany and Poland and overseen by the Russians, that returned the presidency to its 2004 powers and required fresh elections. Although opposition parties also accepted the terms, the protestors did not. Only Yanukovych's immediate resignation would satisfy them now. Uncertain about what to do next, morale amongst the security forces collapsed. Officers simply abandoned their posts and went home.

The game was up. Yanukovych fled, first to Kharkiv and then to Russia via Crimea. Evidence later revealed his preparations had begun a few days earlier. The president's parliamentary support collapsed too. Many government MPs left Kyiv. Those who remained switched sides. Sitting in emergency session, the Rada voted to remove the president with immediate effect, appoint its speaker as interim president, and call immediate elections. This was a move not entirely constitutional. Article 111 of the constitution grants parliament authority to remove the president should the incumbent be guilty of high treason or other crime but required a commission of inquiry first. The Rada skipped the investigation and proceeded straight to the removal, the speaker justifying that on the grounds of the president having already violated the constitution by abandoning his post and by the urgent necessity to restore government that this created. Three hundred and twenty-eight MPs voted to remove the president, and 116 were either not present or voted against.

Putin denounced what he described as a Western inspired coup against a legitimate president. It was the final days of the Sochi Olympics, an extravagant celebration of Putin's Russia. That his moment in the limelight was being overshadowed by events in Kyiv only magnified Putin's annoyance.

Crimea

According to Putin, the decision to invade and annex Crimea was taken in the closing days of the Sochi Olympics. The Olympics ended on 23 February; the day Yanukovych fled. This suggests the decision may have been taken before that, hinting that Putin knew Yanukovych intended to flee. It is commonly assumed that this was a slip of the tongue, that

Map 4.1 Ukraine

Source: iStock/PeterHermesFurian.

the decision was taken *as* the Olympics ended and Yanukovych fled. Either way, it was a decision taken very quickly, suggesting forethought. Evidence for that can also be found in what happened in the days following Putin's decision. It is clear to any observer that the Russian military had planned and prepared for this eventuality. Most likely, the invasion of Crimea was prepared as a contingency for Ukraine's 2015 presidential election should Yanukovych have lost or faced another Orange-style revolution. The central point is that like the invasion of Georgia, the invasion of Crimea was not an entirely spontaneous reaction to events. The decision was taken by a small group of the president's closest associates that didn't include foreign minister Sergei Lavrov. Within that group, only defence minister Sergei Shoigu was reticent. Others, like Nikolai Patrushev, head of the Security Council, and Sergei Ivanov, head of Presidential Administration were strongly in favour. Outside the immediate decision-making circle, close Putin allies like Rosneft director Igor Sechin were strongly supportive too. Putin is not a lone wolf. He sits at the centre of a pack of like-minded people supported by a cast of thousands who think very much as he does.

From what we can tell, Putin's objectives seem to have been threefold. First, to permanently inhibit Ukraine's capacity to associate with the EU and NATO by seizing territory. Second, to establish political leverage over Kyiv either by plunging the government into chaos, the state into bankruptcy, forcing Yanukovych's return, or some combination of the three. Third, to accomplish the Russian nationalist dream of annexing Crimea, thus solving the Black Sea fleet problem, returning "sacred" territory to the motherland, and turning political disaster into a victory sure to burnish Putin's place in Russian history and the public's opinion.

After a few days of relatively minor political agitation in which pro- and anti-Maidan protestors clashed, on 27 February, armed fighters bearing no insignias seized the Crimean parliament in Simferopol and other government buildings and raised Russian flags. Presumably under some duress since there were armed men in the room, the parliament voted to remove Crimea's prime minister and install the pro-Russian Sergei Aksyonov, a shady figure with links to organized crime. The new Crimean government called for an immediate referendum on union with Russia. Berkut units recently disbanded by the Rada in Kyiv took control of crossings between Crimea and the rest of Ukraine. The next day, unmarked soldiers – Putin's "little green men" – appeared across Crimea. They seized airports and government buildings. They surrounded the few Ukrainian bases on the peninsular. The Russian navy meanwhile blockaded the small Ukrainian fleet. Within a few days the little green men had control of most of the peninsular, helped by meticulous planning and the fact the Ukrainians put up no resistance. Many Ukrainian troops swapped sides, those that didn't were instructed by Kyiv to not resist. The Kremlin insisted the little green men were not Russian soldiers but Russian Crimeans, rising spontaneously against an illegal Kyiv government. That was clearly nonsense.

Crimea's referendum, held on 16 March, duly delivered the result Putin wanted. The annexation was rubber stamped by the Duma the next day. Crimea was Russian once again. To mark the occasion, Putin delivered a set piece address to the Duma and Federation Council. What he said bears paying attention to because it pulled the main threads of Putinism together. Putin opened with a direct appeal to imperialism. Crimea is Russian. Always has been, always will be. Its separation from Russia was a great historical injustice. "In people's hearts and minds, Crimea has always been an inseparable part of Russia. This firm conviction is based on truth and justice and was passed from generation to generation, over time, under any circumstances". Next, Putin emphasized the immense threat posed by Ukraine. Not by its turn towards Europe but by its violent coup of "nationalists, neo-Nazis, Russophobes, and anti-Semites" which imperilled the country's Russian-speakers. Finally, he connected this threat to Russia's global struggle against the West. "Like a mirror, the situation in Ukraine reflects what is going on and what has been happening in the world over the past several decades". The West, he insisted, believes it can decide the "destinies of the world", that only it "can ever be right". Its principal weapon in this war, "a whole series of controlled 'colour' revolutions" which produced only chaos and violence. Russia was the West's primary foe. "They are constantly trying to sweep us into a corner because we have an independent position, because we maintain it and because we call things like they are and do not engage in hypocrisy. But there is a limit to everything. And with Ukraine, our western partners have crossed the line, playing the bear and acting irresponsibly and unprofessionally". Some MPs wept with joy. Later, Putin admitted what everybody already knew: the "little green men" were Russian soldiers.

Crimea's annexation came as a huge shock to the West. Not since the Second World War had a major power so blatantly invaded and annexed part of a neighbouring state. The sense of shock first produced inertia as governments tried to make sense of things and realign their priorities. Western outrage was therefore not matched with robust policy. It was also balanced by concern about potential escalation. Putin threatened a nuclear response if Western governments stepped in to help Ukraine. Fearful of any escalation, the West advised Kyiv to exhibit restraint too. The UN General Assembly passed a non-binding resolution rejecting Crimea's annexation, but although 100 states voted in favour, 58 abstained, suggesting that although few supported Russia's actions, a sizable number of states valued their relationship with Moscow more than they valued Ukraine's sovereignty. The US and EU imposed only limited sanctions on Russia targeting a few individuals and discrete sectors – a show of disapproval more than a genuine attempt to weaken, deter, or coerce.

Donbas

Crudely put, post-independence, central and western Ukraine tended to be more nationalist, liberal, and Western oriented, the east more ambivalent about nationalism, more conservative, and more Russian oriented. Most people in the east spoke Russian, not Ukrainian, as their native tongue. But that doesn't mean Ukraine was divided ethnically. Most Russian-speakers, among them past president Leonid Kuchma and future president Volodomyr Zelensky, consider themselves Ukrainian, not Russian, and speak both languages with varying degrees of fluency. Nor does it mean that most people in the eastern Donbas region wanted to secede and join Russia. A large majority there had voted for independence in 1991 and polls showed only modest support for secession even after Euromaidan. This fact – that most people in Donbas wanted to remain within Ukraine – is central to understanding what happened there during and after the annexation of Crimea. Putin's Russia orchestrated an armed uprising to cripple Kyiv and siphon off territory, assuming things would pan out as they had in Crimea. But the Kremlin overestimated its support in Donbas and armed insurrection failed to provoke a popular revolt. Meanwhile, Ukraine succeeded in reorganizing itself. Elections brought in a new president and sense of national purpose. The armed forces regrouped and moved against the insurrection which would have likely collapsed entirely were it not saved from defeat by Russian military intervention. War in Donbas claimed 10,000 lives and divided Ukraine but for Putin the results were almost wholly counterproductive.

The uprising began like Crimea's. In March and April, small groups of anti-Maidan activists occupied government buildings across eastern and southern Ukraine, including in Kharkiv, Mariupol, Donetsk, Luhansk and Sloviansk. Tapes later released by the Ukrainians exposed Glazyev as the orchestrator. The strategy was like that employed in Crimea: use local proxies to seize territory, legitimize separation from Kyiv by hasty referendum, then supply overt Russian assistance. By these means, Russia hoped to fragment Ukraine, give itself leverage, and keep Ukraine permanently out of the EU and NATO.

However, Russia's Donbas operation was a hasty improvization, its basic assumptions wrong. Security forces did not switch sides en masse as they had in Crimea. The separatists did not inspire widespread popular support. They were ejected from Kharkiv's government buildings after just a few days and suffered a similar fate in Mariupol, whilst in Odessa pro- and anti-Maidan protestors clashed violently, resulting in more than 40 deaths in early May when a building being used by the latter was torched. A more direct approach was taken in Sloviansk. In May, the city was seized by Igor Girkin (*nom de guerre*, "Strelkov" – "shooter"), a former Russian soldier and FSB officer turned freelance imperialist, whose battle credits included fighting Moldovans in Transnistria, aiding the genocidal Bosnian Serbs in Bosnia, supressing Chechens, and most recently Crimea, and a band of 60 armed men. The Kremlin alleged these were entirely spontaneous local uprisings against Euromaidan but they were nothing of the sort. It seems likely that Girkin and his band were encouraged if not outright invited to move by Glazyrev, Putin's Donbas organizer. The separatists also held out in Donetsk and Luhansk, where they faced little resistance (largely because no loyalist forces were stationed nearby). Referendums were organized, their results fabricated. Girkin proclaimed himself Supreme Commander of the Donetsk Peoples Republic (DNR) and with its twin, the Luhansk People's Republic (LNR), declared independence. Neither, however, enjoyed majority support in Donbas, nor was either capable of self-sufficiency.

By late May, the government in Kyiv was starting to get its act together. Chocolate tycoon Petro Poroshenko comfortably prevailed in a free and fair presidential election, and Ukrainians rallied behind him. Polls show support for the government surged as did antipathy towards Russia, even in the east where only half expressed positive feelings towards Russia. The Ukrainian army, capable of deploying only around 6,000 soldiers, was fortified by volunteers, private militia and other armed groups including the far-right Azov Brigade which was later incorporated into the regular army and took to the offensive. The Ukrainians retook Donetsk airport on 26 June and, after a bitter fight that left more than 150 people dead, Sloviansk a couple of weeks later.

Russia had failed to ignite a general insurrection in the east and had misjudged Kyiv's capacity to reconstitute itself. Its proxies in full retreat, the Russians poured weapons, including tanks, GRAD missiles, and anti-aircraft systems, into Ukraine, likely hoping to engineer a stalemate like those achieved in Abkhazia and South Ossetia. The fighting intensified. Ukraine used fighter jets against separatist positions. The separatists shot down a troop transporter approaching Luhansk, killing 49.

The West gathered itself too. As with Crimea, the initial reactions of figures like US president Barack Obama and German chancellor Angela Merkel were shock, concern and a focus on preventing escalation. By the time Poroshenko's government was able to consolidate Ukraine's response, however, the West's position had hardened too. The annexation of Crimea represented a blatant violation of the most basic rules of international law, one that harked back to a time when imperialists gobbled up the weak with impunity. No matter what the Kremlin claimed, it was clear Russia was expanding its aggression into Donbas. To be sure, some Europeans (in Germany especially) were reluctant to step up pressure on Russia, fearing the economic pain tougher measures would cause them, but the EU and US widened and deepened economic and political sanctions targeting more sections of the Russian economy, oligarchs close to Putin, and Russia's global political standing. The EU also rapidly concluded its association agreement with Ukraine, while the G8 expelled Russia, becoming the G7 once again. Putin and Medvedev reacted with shrill threats of nuclear retaliation. Russian military aircraft tried to stoke fears by conducting illegal flights over Estonia with increasing regularity. Russian submarines appeared in Swedish waters. All attempts to bully and deter.

On 17 July 2014, a Russian surface-to-air missile destroyed Malaysian Airlines flight MH17 as it flew high above Ukraine, killing all 283 people aboard. The Kremlin denied involvement, blaming everyone but those responsible. It claimed the plane was shot down by a Ukrainian fighter jet. It also claimed the affair was orchestrated by the British as a "false flag", a plane packed with corpses before being detonated over Ukraine. The truth, however, was confirmed by Dutch courts in 2022 which found that a Russian Buk anti-aircraft battery had been sent into Ukraine and was employed by the separatists, with Russian supervision. Thinking they had detected a military transport aircraft, they fired a missile which hit its mark on the Malaysian airliner. The mass killing of innocents and the breathtaking cynicism of Russian denials angered Western public and government opinion alike. Perhaps for the first time, Western leaders began to see Putin's Russia for what it had become. Sanctions were widened to include the energy sector, while other measures focused more explicitly on countering Russian assistance to its proxies.

Ukrainian government forces retook Ilovaisk in mid-August, the furthest extent of their counter-offensive. Putin now faced choosing between letting his proxies fail or sacrificing the veneer of distance by intervening directly. He chose the latter. The Russian army began deploying regular soldiers into Ukraine and conducting combat operations themselves. As in Crimea, Russian soldiers bore no insignia and were regularly rotated to maintain the fiction that there were no Russian forces inside Ukraine. Their iPhones and Fitbits told a different story, however. These devices allowed anyone who cared to check to place thousands of Russians troops squarely inside Ukraine. The DNR's "prime minister", Alexander Zakharchenko, boasted that Russian troops and tanks were now fully engaged – although he later retracted and followed the Kremlin's line that some Russian soldiers had been so moved by the plight of Donbas they had taken holidays to volunteer to fight there. Russian intervention shored up the separatists' military position. Ukrainian forces in Ilovaisk were surrounded and destroyed, killing more than 350 soldiers. Ukraine lost another 700 wounded, missing, or taken prisoner, a catastrophic loss for such a small force which provoked strong criticism of the government's strategy. Luhansk airport, which had become a major Ukrainian base, was destroyed by withering artillery fire and Donetsk airport came under sustained attack, although the defenders held out. But although Russian intervention had stabilized the situation, but it would take a much greater level of force to defeat the Ukrainians. That would eviscerate whatever was left of Russia's claim to not be directly involved. With international criticism mounting and Western sanctions escalating, this was not a step Putin was prepared – yet – to take. Deeper intervention might jeopardize Russian relations beyond the West, including with China, and draw Russian forces into a costly occupation from which it would be difficult to withdraw.

Putin turned to diplomacy to consolidate his gains and find a pathway out of sanctions. Poroshenko was also ready to do a deal. Although Ukraine's immediate situation had stabilized, its heavy defeat at Ilovaisk, and the increased pressure on Luhansk and Donetsk airports hinted at prolonged vulnerability. Ukraine had diminishing numbers of deployable forces and limited control over the private militias fighting on its side. In direct firefights with the Russians, Ukrainian forces typically came off second best. Meanwhile, Ukraine's economy, troubled at the start of the crisis, was heading towards catastrophe. Emboldened by

Russian support, the separatists saw little reason to negotiate. Having let the genie out of the bottle, the Kremlin found it difficult to put it back in. Yet neither could Putin abandon them without being accused of betrayal. The result was an ambiguous compromise. Mediated by Merkel and French president François Hollande, the "Minsk I" protocol agreed on 5 September 2014 provided for a ceasefire, monitored by the OSCE (European governments, especially Germany, rebuffed Ukrainian pleas that the EU or UN be given responsibility for peacekeeping). All foreign (i.e., Russian) forces were to be withdrawn and OSCE observers stationed to monitor the Ukraine/Russian border. Donbas would be granted special status, with civil authority decentralized and local elections held under Ukrainian authority to select a new regional government. The deal thus gave all sides something of what they wanted. The Russians also benefitted from the fact it made no mention of Crimea. But the arrangement was riddled with inconsistencies, not least the fact it left the difficult question of sequencing unresolved. Ukraine argued that Russian forces be withdrawn, OSCE border posts installed, and central authority established prior to elections and constitutional reform. The Kremlin and separatists insisted it happen the other way around. Part of the problem was that the Russians and separatists had unfinished military business since they did not actually hold anything like all of Donetsk and Luhansk oblasts. In effect, therefore, Minsk I provided political cover for a strategic pause and little else.

Russia used the pause to strengthen its position inside Ukraine. Emboldened, in the new year the separatists launched a new offensive. Donetsk airport bore the brunt and fell in January. The battle for Debaltseve, a government-held saliant northeast of Donetsk, was even bloodier. A large Ukrainian force held on for weeks as separatist and Russian forces surrounded and bombarded the town. Only when access to the town was almost completely cut did the Ukrainians finally retreat, not before they had lost more than 250 soldiers with more than 600 captured and wounded. The separatists lost around 800 fighters killed and injured. Russia also lost a significant number of soldiers, with as many as 70 killed according to some Russian sources but likely more than that. They had improved their position but had not achieved a decisive breakthrough.

A new deal ("Minsk II") was struck in February. In essentials, the second Minsk deal resembled the first but it pushed a little further on the constitutional reforms required of Ukraine and clarified details with respect to the ceasefire. The political ambiguities remained, however, and military analysts in the West and Ukraine warned that Russia and the separatists were gearing up for another offensive. Sure enough, in June 2015, DNR separatists launched an offensive on Marinka, likely a test of Ukrainian defences and resolve. The attack was repelled by a determined defence that cost the separatists more than 200 dead and wounded. Defeat at Marinka seemed to convince Moscow that further territorial gains were unlikely unless Russian forces took the lead, a step Putin was unwilling at this stage to take. Although Minsk II's ceasefire never entirely held, military operations on both sides were wound back as both dug in. Meanwhile, no progress was made on the protocol's political aspects.

Consequences

Russian nationalists wept with joy when Crimea was annexed. Putin publicly feted, his flagging popularity transformed. Putinism, reconfigured as more imperialist, nationalist, militarist and authoritarian was back. Or so it seemed. In fact, Russia's invasion of Ukraine had failed to achieve its principal objectives.

From what we can tell, Putin had three aims in mind when he ordered the annexation of Crimea and the war in Donbas: prevent Ukraine associating with the EU, establish leverage over Kyiv by fragmenting Ukraine, and annex Crimea. Russia not only failed to achieve its first two objectives, it put them out of reach. The achievement of Russian objectives in Ukraine depended on the ability of Russian and Russian-backed forces to secure an easy victory and on the West being cowed into only a modest response. The latter requirement explains why Russian strategy hinged on the fiction that these were spontaneous local conflicts. The pretence in Crimea wasn't maintained for long, but then it wasn't needed for long because Russia took control of the peninsular quickly. But in Donbas, Ukraine resistance was much fiercer and local support much lower than the Kremlin had expected. This forced Putin to choose between defeat or relinquishing the fiction of Russian non-intervention. Putin chose the latter. Barely tenable at the outset, the fiction evaporated entirely with the downing of MH17 and Russian military intervention to prop up the ailing Donbas separatists. Intervention saved DNR and LNR from defeat but denied Russia its main objective.

Instead, Russia acquired responsibility for three restive and dependent territories. The Kremlin struggled to control its local proxies, especially as those in eastern Ukraine grew resentful of what they saw as Moscow's limited support. Mercenaries were brought in to forcibly disband some recalcitrant brigades and it took a brutal campaign of assassinations in 2015 to force the local leaderships to toe Putin's line. Meanwhile, cut off from market access, power supplies and fresh water from Ukraine, Crimea was not economically viable. Estimates suggest that just maintaining it cost the Kremlin up to $5 billion each year. That was in addition to the $3.5 billion cost of constructing a 17 km-long bridge over the Kerch Strait to connect Crimea to Russia. Costs were similarly high in the Donbas, where in addition to propping up a collapsed economy, defences and military operations also needed sustaining. With no prospect either would achieve international recognition, these costs would become a permanent burden.

Since his Munich speech at least, Putin had imagined himself locked in geopolitical struggle with the West. It had been a struggle of his own imagining since there was in fact no Western conspiracy of support for colour revolutions or active competition with Russia for global influence. In the West, there was antipathy and division not enthusiasm about NATO enlargement; Western governments were by and large reducing not increasing military spending; most saw their primary threat in radical Islamism, Russia figured little or not at all in defence planning. Russia's invasion of Ukraine changed some of this. It forced NATO to recognize the threat Russia posed and rethink its priorities. Western governments condemned Russian aggression, imposed albeit initially quite limited sanctions, and rallied in support of Ukraine. The EU concluded the association agreement that lay at the root of the dispute. Nervous eastern members demanded NATO reprioritize and their allies at last agreed. The alliance's Rapid Response Force was increased to 40,000, a 5,000-strong enhanced forward presence was established in Estonia (led by the UK), Latvia (led by Canada), Lithuania (led by Germany) and Poland (led by the US) with smaller deployments in Bulgaria, Hungary, Romania and Slovakia. The alliance increased air and naval patrols. This represented a significant realignment of the alliance's strategic focus and substantial increase in its military presence near Russia's border. Thus by his aggression in Ukraine, Putin conjured up precisely what he claimed to fear most. Russian aggression had pulled Ukraine and the West closer together.

Russia's global standing was harmed too. It lost its membership of the G8 and was subject to a series of negative votes in the UN General Assembly. Only a tiny handful of states supported Moscow; a very large majority condemned it outright. Major non-Western powers like China and India refused to condemn Russia, but they also refused to support it, leaving the Kremlin more isolated and more dependent on others. The invasion also had a chilling effect on prospective members of the Eurasian Union: Moldova, Tajikistan and Azerbaijan all went cold on the idea, while Ukraine joined Georgia and the Baltic states in the list of former Soviet republics now squarely outside the *Russkiy mir*.

Further reading

Paul D'Anieri, *Ukraine and Russia: From Civilized Divorce to Civil War*. Cambridge: Cambridge University Press, 2019.

Lawrence Freedman, *Ukraine and the Art of Strategy*. Oxford: Oxford University Press, 2019.

Robert Horvath, *Putin's Preventive Counter-Revolution: Post-Soviet Authoritarianism and the Spectre of Velvet Revolution*. Abingdon: Routledge, 2015.

Serhii Plokhy, *The Gates of Europe: A History of Ukraine*. London: Penguin, 2015.

Serhii Plokhy, *Lost Kingdom: A History of Russian Nationalism from Ivan the Great to Vladimir Putin*. London: Penguin, 2017.

Anna Reid, *Borderland: A Journey Through the History of Ukraine*. New York: Basic Books, 2015.

5
Syria

In early 2011, three years before Euromaidan, the Middle East was convulsed by popular uprisings, the Arab Spring. First, Tunisia's long-serving president Zine El Abidine Ben Ali was driven from power. Then, Egypt's Hosni Mubarak, another long-serving authoritarian, was ousted by protestors. The uprising convulsed Libya too, but Muammar Gaddafi turned his guns on peaceful protestors. When the protestors fought back, the country slid into civil war. Prompted by pressure from the Arab world and Europe, the UN Security Council authorized a NATO-led intervention to protect Libyan civilians. Russia abstained in that vote, a bone of contention between its president, Dmitry Medvedev, and prime minister, Putin. Oman, Saudi Arabia, Jordan, Bahrain and Yemen also experienced mass uprisings. The first three used limited political reform and economic incentives to mollify the crowds; the government in Bahrain restored order violently with the help of Saudi military intervention; Yemen careered into civil war.

It was Syria, however, that loomed largest. The uprising there led to a bloody civil war, the rise of the radical jihadist Islamic State (IS), and drew in Iran, Hezbollah, the US, the UK, France, Saudi Arabia, Qatar, Jordan, Turkey and Russia.

Before Crimea, Medvedev and then Putin had steered a course focused on preventing Bashar al-Assad's violent overthrow without direct military intervention. In autumn 2015, however, Putin ordered the Russian military into Syria to prop-up Assad's flagging regime, his position hardened by three considerations: Western timidity which created a vacuum he was happy to fill; fear that Assad faced imminent defeat; and confidence in the capacity of Russian military force to achieve its political goals of defending Assad, extending Russian influence into the Middle East, and forcing the West to acknowledge Russia's great power status by making itself indispensable to peace in Syria. Taking advantage of Western timidity, disunity and confusion, Putin ended Russia's brief post-Crimea diplomatic isolation by positioning itself as the great and indispensable power necessary for peace in Syria. But victory in Syria came at a terrible price: a civil war that consumed more than half a million lives and displaced more than half Syria's pre-war population. The Syrian government and its allies were responsible for around 90 per cent of all civilian deaths, Russian forces alone killed at least 7,000 Syria civilians.

Before the intervention

Syria was important to Moscow, but it was not Russia's only regional concern. Putin worked hard during his first two terms to improve relations with Israel, Saudi Arabia, Qatar, Jordan, the UAE and – above all – Turkey, establishing commercial and military ties with them all. The Soviet Union had enjoyed some influence in the Middle East and with the US-led War on Terror not just floundering but causing serious political pushback, it was precisely the sort of place likely to prove receptive to Putin's vision of a multipolar world. Thus, when the crisis in Syria erupted in March 2011, the Kremlin hoped to avoid having to choose between Damascus and these hard-won relationships. Like almost everyone else, Putin believed Assad could be persuaded to show restraint and mollify the protestors with limited reform. Ultimately, however, he was forced to choose between the government and the people, and he chose Assad.

From the start, the Kremlin saw Syria's crisis as a national security and geopolitical problem, not a human rights issue. Damascus was a longstanding ally, Russia's last in the Middle East. Cooperation with Syrian intelligence was critical to Russian knowledge and influence in the region. The Russians maintained listening stations in Syria, keeping tabs on Israel, Lebanon and the Americans in Iraq. Russia also retained a decrepit naval resupply base at Tartus, its only base in the Mediterranean (refurbished after 2012). The two countries were also trading partners, and Syria was a major client for Russian arms.

The Kremlin viewed the Arab Spring as a worrying extension of the colour revolutions and Putin saw American hands behind it. He also saw a rising tide of regime change unilateralism behind the US invasion of Iraq in 2003 and NATO-led intervention in Libya in 2011. It was not that Russia cherished the sovereignty of others. In Putin's worldview, only great powers enjoyed true sovereignty; the sovereignty of lesser states was more ambivalent. What worried him was the extension of *Western* power through the unseating of friendly governments. Putin feared and disliked popular protest in equal measure for, in his view, it led to fragmentation, chaos and decay. He also struggled to conceive that peoples might think and act for themselves quite independently of geopolitics. Russians welcomed the Kremlin's vocal opposition to the West, they applauded the defence of Syria's Orthodox Christians and other minorities from Islamic extremism. When it came, Putin's war for Assad was blessed by the Russian Orthodox Church. Patriarch Kirill called it a holy war.

The Syrian government reacted to the outbreak of protests in March 2011 with violence. Government forces fired on unarmed protestors and detained en masse those suspected of supporting the opposition. Many of those detained were tortured; at least 40,000 Syrians were tortured to death in the first few years of the civil war. Putin thought Assad's heavy-handed response courted disaster but also wanted to protect the Syrian regime from any international pressure that could undo it. Thus, with one hand, Russia tried to protect Assad from sanctions and intervention, whilst on the other trying to persuade him to change course. This strategy proved self-defeating since the more Russia objected to international action, the more it encouraged Assad to use violence not reform to keep himself in power. The problem was compounded by Moscow's inability to control Syria's president. Influence is not a one-way street. Just as the Kremlin could apply pressure on Damascus, so Damascus had leverage over the Kremlin since Russia's interests depended on Assad's political survival. Over time, Assad came to understand and employ his leverage to good effect. Meanwhile, Putin's ability to influence Assad was limited: he had a sledgehammer to crack a walnut but desperately wanted the nut kept intact.

Russia used its veto to prevent the UN Security Council imposing sanctions or other measures to persuade Assad to relent. It did so even when the Syrian leader walked away from a 2012 peace plan composed by former UN Secretary-General Kofi Annan that Russia had previously backed. By inhibiting international pressure, Russia helped tip

Syria from violent protests to all-out war. By blocking multilateralism, it forced the West to think unilaterally. One consequence was that US president Barack Obama called for Assad to step aside, a symbolic act that Obama had no intention of following through with, but which nonetheless set the terms of the looming clash. Tripoli fell to Qatari-backed rebels in Libya just two days after Obama's call, catching Western governments and the Kremlin alike by surprise. Gaddafi was tracked down and viciously killed by rebels that October. All this confirmed Putin's sense that Syria was another stage in a broader geopolitical struggle between the West and the rest.

The NATO-led intervention in Libya succeeded in ousting Gaddafi but did not bring peace. On 11 September 2012, the eleventh anniversary of the 9/11 atrocities, the US embassy in Benghazi was attacked by jihadists and the ambassador and three other officials were killed. The Benghazi murders caused bitter recriminations on Capitol Hill that tied up the White House for years. Obama's Libya policy was castigated for causing the problem in the first place. Things were not helped by Libya's descent into chaos as its divided parliament failed to establish authority whilst on the streets armed factions and jihadists jostled for control. Meanwhile, in July 2013, Egypt's first democratically elected government, led by Mohammed Morsi and the Muslim Brotherhood, was overturned by a military coup. Headed by Abdel Fattah El-Sisi – the former head of military intelligence now serving as defence minister – the new military government conducted a violent crackdown on its opponents, gunning down between 800 and 2,500 unarmed protestors. Before Crimea, then, chaos in Libya, democratic reversal in Egypt, and the West's evident lack of commitment to democracy in either place had already emboldened Putin. It is one of Putinism's many contradictions that it sets itself against an aggressively hegemonic West that it also sees as being corrupted, divided and weak.

Putin was still unsure, however, about American intentions towards Syria. Obama had publicly declared that any use of chemical weapons by the Assad regime would be a "red line", something which everyone assumed meant would trigger US intervention. The decisive moment came on 21 August 2013 when government forces dropped sarin gas on eastern Ghouta, just outside Damascus. Approximately 1,400 people were killed, and four times that number injured, including more than 400 children. Russian officials, Putin among them, insisted there was no evidence of Syrian government responsibility, a point from which they never demurred despite the UN proving it beyond doubt. They suggested instead that it was a false flag to provoke Western intervention. Russian diplomats paraded captured Syrians who faithfully recounted that there had been no chemical attacks. Putin, of course, knew full well that Assad's forces were behind the chemical attacks, but he could not control the Syrians and feared Western intervention. He took the unusual step of expressing his opposition to Western intervention in an op-ed penned for the *New York Times*.

With Obama having declared chemical weapons a "red line", US intervention was widely anticipated. But with plans approved and missiles ready to fly, the president got cold feet. He had come to power promising to rid the US of its costly military entanglements in the Middle East, not take them into new ones. Libya had not turned out as he had hoped, and he feared a slippery slope in Syria. At the eleventh hour, Obama pulled back from the brink and announced his intention to seek congressional approval. It was clear from the start, however, that he did not have sufficient support there. Meanwhile, desperate to prevent US intervention, Sergei Lavrov indicated to his American counterpart John Kerry that Russia might be open to a deal to peacefully disarm Syria's chemical weapons. Obama raised it directly with Putin at a G20 Summit in Saint Petersburg and Putin agreed, saying he would persuade Assad to accept the deal. The plan was quickly endorsed by the UN Security Council, and inspectors dispatched to identify, ship and destroy Syria's chemical weapons. The deal allowed Obama to claim he had delivered on his "red line" without using force; it worked for Putin by taking US military intervention off the table. But in practice, the arrangement neither disarmed Syria of its chemical weapons nor deterred their future use. Damascus lied, obfuscated and confused the inspectors and thereby retained its chemical weapons capability. It soon began using them again. Chlorine bombs were used within 200 days of the disarmament deal, sarin perhaps as early as 2016 and certainly by early 2017. Once the threat of American intervention had passed, conventional bombs fell indiscriminately with even greater frequency than before.

The chemical weapons deal had hinged on the US threat of force. Putin pursued it only because he judged it the only way of preventing US intervention. We now know that the package of strikes Obama approved was modest but the Russians at the time did not know that.

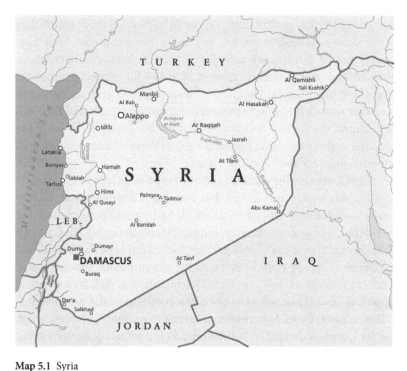

Map 5.1 Syria

Source: iStock/PeterHermesFurian.

There was every reason for Putin to worry that once intervention began the US would pursue things to their bitter end as it had in Libya. Once the threat of intervention receded, the diplomatic leverage it had created evaporated. Stepping back may have been the prudent thing for the US to do, but it also conveyed weakness to a Russian president who measured world politics in terms of power. By backing down, the US handed the initiative to the Kremlin and not just in Syria. American weakness undoubtedly encouraged Putin's boldness when the Euromaidan crisis erupted less than four months later.

Meanwhile, things got worse in Syria. In June 2014, an extremist group calling itself the Islamic State (IS) over-ran Iraq's second city, Mosul. When they took the city's Badush prison, the jihadists separated Sunni and Shi'ite prisoners, forced more than 600 Shi'ites to kneel beside a ravine and shot them. Other massacres followed. IS encircled Tikrit and hundreds of fleeing soldiers were captured and more than

700 executed. At Camp Speicher, more than 1,000 Shi'ite air cadets were captured and executed. As they cut a bloody swathe across Iraq, they declared a new caliphate under the leadership of Abu Bakr al-Baghdadi. The Yazidis suffered most of all, victims of a genocide. IS had bombed the Yazidi towns of Til Ezer and Siba Sheikh Khidir in 2007, killing almost 800 people. Now, they seized the Yazidi majority town of Sinjar and unleashed hell. Yazidi men and older boys were rounded up and executed, while thousands of Yazidi women and girls were kidnapped into sexual slavery. Women deemed too old or ugly for sex were killed and buried in mass graves. The battered and traumatized survivors fled to the barren Mount Sinjar. Iraq was splintered in less than a week, the state propped up by Kurdish Peshmerga in the north and Shi'ite militia in the south. The US led a coalition to intervene from the air to push IS back, the ground fighting done by Iraqis themselves – albeit with American help.

The chaos quickly spread into Syria. IS with recently captured arms, took Raqqa and advanced south and west. Those who resisted suffered a terrible fate. The al-Shaitat tribe fought back. When they too were defeated, IS fighters shot, beheaded and crucified at least 700 of them. The extremists also unleashed a wave of terrorism on Europe, among the most horrific of the dozens of attacks was the November 2015 atrocities on Paris which left more than 130 people dead. Meanwhile, around one million refugees poured out of Syria seeking refuge in Europe. The dramatic rise of IS coincided with the war in Donbas. Once the dust settled from that, Putin surveyed a very different geopolitical landscape and recalibrated Russia's policy in Syria.

Intervention

War in Ukraine loomed large over the November 2014 G20 meeting in Brisbane, Australia, especially the downing of flight MH17. Russia's diplomatic isolation reached its nadir in Brisbane, Putin was publicly humiliated. Australia had lost 38 citizens aboard MH17 and its prime minister, Tony Abbott, promised to "shirt-front" the Russian leader when he arrived – a colloquialism for hitting someone in the chest drawn from Australian rules football. There was no shirt-fronting but there was plenty of embarrassment for Putin. At the group photo, Putin was placed not in his customary position in the centre with Obama and Chinese leader Xi Jinping, but on the edge next to South Africa's Jacob Zuma. At the leaders' lunch, he was shown to a table accompanied only by Brazil's Dilma Rousseff. Leader after leader either ignored or rebuffed him. Putin was angered not chastened by this experience. He left the summit early, resolved to not accept any future invitations to visit the West until he – and Russia – could be sure of receiving the deference he believed deserved. What looked like a tall order at the time, took less than 12 months to achieve. War played a central role.

Putin's self-imposed exile ended when he visited New York in September 2015 to address the UN General Assembly. There, he explained that Assad was waging a valiant fight against terrorism and that Russia intended to support him. Barely 72 hours later, a Russian general walked into the US embassy in Baghdad and announced that airstrikes would begin in Syria within the hour. An hour after that, Russian bombs struck Syrian opposition targets in Hama and Homs.

Putin had been considering military intervention for months, troubled by assessments that Assad's government was in danger. Despite extensive Iranian help, Syria's military was stretched thin, its chronic manpower shortage becoming acute. Syria's opposition, meanwhile, was getting more capable. It scored a dramatic victory in May 2015 by seizing Idlib after a short offensive. A second offensive in Daraa in the south failed, but the Kremlin judged the long-term trends tilting in the opposition's favour and believed Assad's defeat likely unless something was done to alter the balance of forces. The risk Assad's fall might present was underscored when the "Caucuses Emirate", the largest jihadist insurgency in Russia, pledged allegiance to IS when then claimed responsibility for two terrorist attacks in Dagestan. In the summer of 2015, Iran's supreme leader Ali Khamenei despatched a senior envoy to appeal for Russian help. Putin agreed and Russian officers and defence officials spent the summer planning with Qasem Suleimani, head of the Iranian Republican Guards' Quds force. Their strategy involved using Russian air power to help the Syrian government take the initiative in Aleppo, Hama, Daraa, Latakia and Palmyra. Battlefield losses exacted from the air would force the opposition to concede to Assad's terms. With the West in retreat, Putin sensed an opportunity to plug the great power vacuum and make Russia the indispensable power in Syria. Russian generals spoke of establishing permanent bases there.

The plan appealed to Putin immensely. It called for a Western-style military operation, a demonstration of Russian military supremacy from the air in which Russian forces assumed few risks while Putin pulled political strings. Syria signed over control of Khmeimim air base giving Russia a second Middle Eastern base: the hallmark of a great power. Significant work was done to improve the runway and facilities. Two thousand Russian troops, including elite marines, were deployed with initially 24 aircraft, supported by tanks, helicopters and anti-aircraft weapons. The number of aircraft based in Syria grew to 44 by 2016, augmented by Cold War-era long-range Tu-160 and Tu-95MS bombers operating from bases in the Caucuses. The ground force grew to around 4,000, augmented by up to 5,000 mercenaries employed by Wagner Group, a private military company with close links to the Kremlin. The mercenaries included several thousand Chechens. A further 1,700 troops were deployed at the naval base in Tartus. Around 15 warships mainly from Russia's Black Sea fleet took up position off the Syrian coast and warships in the Caspian Sea also set their sights on Syria. Missiles from the Caspian hit targets in Raqqa and Idlib.

Military intervention was intended to protect Assad and end Russia's post-Crimea isolation by forcing the West and the UN to deal with it on Syria. This cannot be emphasized enough, since for Putin ending Russia's diplomatic isolation was a necessary first step to legitimizing its annexation of Crimea. Even if Western powers could not be persuaded to formally recognize that annexation, if they could be forced to treat Russia with reverence, annexation would over time become an irreversible political fact just like the separation of Abkhazia and South Ossetia from Georgia and Transnistria from Moldova. Formal recognition was ideal but not essential. What mattered most to Putin were political facts.

Putin reassured Russians that this would be a limited, low-risk, war against IS terrorists. This would be no Chechnya-like grind. But it was not risk free. In October, IS affiliates bombed a Russian passenger plane over Egypt's Sinai Peninsula, killing 224 people, mostly Russians. In November, a Turkish F-16 shot down a Russian Su-24 it said had violated its airspace. A furious Putin imposed sanctions on Ankara. Bilateral relations between the two soured, undoing more than a decade of diplomatic bridge-building.

In truth, however, Russia's war was not directed primarily at IS, as the Kremlin claimed. This was made plain by its targeting. The principal early targets were mainstream opposition groups, especially those receiving US support. In the first few hours, Russian aircraft attacked al-Lataminah in northern Hama three times, targeting Tajammu al-Aaza, a moderate secular group supplied by the CIA. On 7 October, Russian aircraft attacked a storehouse containing ammunition, artillery, armoured personnel carriers and tanks belonging to Liwa Suqour al-Jabal, another CIA-backed group that was led by defectors from the Syrian army – hardly radical Islamists. Helped by Russian airpower, government forces went on the offensive in Homs, Idlib, Daraa and the surrounds of Damascus – none of them areas where IS was especially active. More than 400 civilians were killed in the first two weeks of October 2015, over half attributed to Russian airstrikes.

As the campaign widened, civilian casualties grew. In November, Russian forces carried out more than 50 strikes on Deir ez-Zor in Syria's east, mainly hitting civilians. More than 70 were killed in a single strike. Evidence of a pattern of indiscriminate force quickly emerged. Russian aircraft bombed a marketplace and residential streets in Idlib, killing more than 40 civilians. In the course of 2015, there were a staggering 112 attacks on medical facilities, 85 per cent of them by government forces and their allies – principally, the Russians. The Russians and their allies employed the strategy of atrocity used in Chechnya – siege, mass indiscriminate destruction, collapse – with the same devastating effects.

Russian intervention gradually turned the tide of battle, but it was slow work. The opposition proved more resilient than expected, the Syrian government more feckless. More important for both Syria and Russia, however, was the immediate political effect. Russian intervention crippled the US strategy of coercing Assad and made Russia the indispensable power for peace in Syria. Leadership, however, came at a price. It was now Putin's job to find a way to end Syria's bloody war.

Plan A: bilateral bargain

Russia needed to translate military gains into political facts. Putin intended to do that by wrapping the peace in an international agreement that included the US and UN. Western recognition was also essential to Putin's wider objectives of asserting Russia's return as a great power, normalizing its diplomatic relations, and thereby strengthening its hold over Crimea. Putin tasked foreign minister Sergei Lavrov with negotiating a bilateral deal with his American counterpart, John Kerry, that could then be sold to the Syrians and the UN. The two pursued a series of national ceasefires to be followed by a political settlement. Bereft of leverage and desperate for a deal, Kerry was willing to countenance a deal that left Assad in power but Assad, emboldened by battlefield success, had little interest in compromise. An exasperated Putin even declared the end of Russia's intervention in a bid to force Assad's acquiescence. Assad dismissed the bluff and insisted Putin choose between backing him and risking Syria's total collapse. Putin backed off and Russian military operations continued and by mid-May 2016, Russian aircraft were at the fore of the battle for Syria's second city, Aleppo. Jets flew from the *Admiral Kuznetsov* carrier in the Mediterranean to increase Russia's sortie rate. Aleppo was fully besieged and opposition-held districts systematically dismantled by air and artillery. Twenty-five hospitals and clinics were destroyed in five months; more than two million people now lacked access to clean water.

As Aleppo burned, the Russians offered a deal. They would call off the assault if the US publicly committed itself to a Russian-led peace process that left Assad in power. Washington said no. The Kremlin offered a more modest alternative: the US agree to direct cooperation with Russian counterterrorism operations, in return for which Russia would urge Assad to ground his air force. Kerry thought that a small price to pay. But it was an obvious trap, a gambit as much about Ukraine as Syria. Putin, recall, wanted to end Russia's diplomatic isolation, formal American military cooperation would help with that but also help Russia achieve its goals in Syria. In return, Putin was promising only to urge Assad, not compel him. The Kremlin would be left to pocket the political boon, without assuming responsibility for what happened in Syria. Besides, the plan covered only Syria's air force, not its ground operations and said nothing about Russian operations. But with little leverage of its own, the US accepted the deal. It would go into effect if a ceasefire held for a week.

In the event, the ceasefire didn't hold. Instead, things unravelled very quickly. On 17 September, with the ceasefire five days old, US and coalition aircraft mistakenly hit Syrian government forces operating in Deir ez-Zor. Putin was furious. Lavrov accused the Americans of mendacity; of using the ceasefire as a cover to attack the Syrian army. Payback came swiftly but against a much softer target than the US military. Syrian helicopters backed by Russian Su-24 jets bombed and strafed a UN and Syrian Red Crescent humanitarian convoy as it unloaded supplies. The vehicles were clearly marked with Red Crescent and UN insignia, their coordinates shared with the Syrian authorities. This was a meticulously planned war crime that killed 14 aid workers and destroyed 18 vehicles. The Kerry–Lavrov process was over.

Russian forces joined the Syrian government's final push on Aleppo. For weeks, Syrian artillery and airpower and Russian aircraft and missiles rained down indiscriminate fire on the city. In the words of the UN's Commission of Inquiry, "in eastern Aleppo, pro-Government forces pummelled vital civilian infrastructure, with disastrous consequences. Day after day, hospitals, markets, water stations, schools and residential buildings were razed to the ground". The civilian death toll soared. Eastern Aleppo's hospitals were bombed out of commission, close to 1,000 civilians killed in this phase of bombing alone. Besieged, starved and smashed, east Aleppo collapsed amidst an orgy of violence. Russia had helped Assad secure a major victory.

In its first 15 months, Russia's war in Syria had reversed the government's fortunes and re-established its own position at the top table of international diplomacy. It was a masterstroke for the Kremlin and a disaster for the West and its allies, the Syrian opposition and Syria's civilians. But it was not all one-way. IS remained a problem and government forces were exhausted and overstretched. Victory in Aleppo, for example, was assisted by a Faustian bargain with Turkey. Turkey agreed to pull out the militia it controlled but in return wanted its own foothold inside Syria as a buffer against the Kurds. Russia achieved its victory in Aleppo but one that was in some respects dependent on Turkey.

Turkey's President Erdoğan had become increasingly fretful of the warming alliance between the US and the Kurds in Syria resulting from the campaign against IS. He had already tried to mend bridges with Putin when an attempted coup against him in July 2016 sealed the deal. While Turkey's NATO allies dithered over whether they wanted the coup to fail, Putin moved decisively. He called Erdoğan, offering Russian military support should he need it, and reportedly authorized the release of intelligence that helped Erdoğan defeat his domestic opponents. Grateful for Putin's help, Erdoğan moved his Syria policy away from alignment with the West and sought a degree of rapprochement with Russia. Ankara and Moscow still had different interests, but their leaders respected one another and played the same game of power politics. In return for concessions to Turkey's security interests, Russia could use Turkey to persuade enough of the Syrian opposition to accept a deal favourable to Assad. That was the foundation for the new plan.

Plan B: Russian-led troika

Donald Trump's election to the White House in November 2016 was a boon for Putin. Trump's personal sycophancy outshone even that of his own acolytes in the Russian media. Trump's Syria policy, meanwhile, veered erratically from the chaotic to the bizarre. When some clarity did emerge, the US limited itself to defeating IS and containing Iran. Human rights and political transition were off the table. That suited Putin well.

Putin's plan B was to circumvent the US and UN entirely. It would come in two phases, Russia at the heart of both. The first involved working with Iran and Turkey to enact a series of conflict freezes around opposition-held pockets. Once the frontline was frozen, the second phase involved summoning the parties to Russia to hammer out a peace deal. Three zones were agreed in advance: Idlib, Rastan/Talbiseh (northern Homs governate and southern Hama), and a southern zone comprising parts of Deraa and Qunaitra governates. A fourth, eastern Ghouta, was more contentious. Turkey pressed hard for its inclusion and threatened to walk out if it were not. The Syrian government obstinately refused – eastern Ghouta was the rebel enclave it wanted to clear next. Putin initially backed Damascus but needed to keep Turkey in the tent so reluctantly agreed with Erdoğan and added eastern Ghouta. Under the plan, ceasefires would be established in each of the zones for a period of six months. The cessation did not apply to IS or other terrorists, effectively licensing ongoing Russian force since it claimed only to be targeting terrorists. Yet despite its many flaws, the exhausted parties agreed. The ceasefires were never fully implemented but violence declined overall throughout 2017.

Next up was Putin's grand summit, held in the glittering pantheon of Putinism, the Olympic resort at Sochi. The "Congress of Syrian

National Dialogue" was the centrepiece of his Syria strategy: 1,500 Syrian delegates were summoned in January 2018 to agree a settlement, one facilitated and guided by Putin's Russia. Putin intended to cajole delegates to accept a largely preselected committee charged with drafting a new constitution for the whole country, one that allowed some Kurdish autonomy, permitted modest power sharing, and provided for presidential and parliamentary elections for which Assad (who would remain in power in the interim) could stand. Meanwhile, the four de-escalation zones would be brought under the authority of the Syrian state. The problem was that Putin's vision was not widely shared by Syrians. Assad was unenthusiastic; the opposition rejected it outright; Turkey baulked too, fearing Kurdish autonomy in Syria would embolden its own separatist Kurds.

So, Putin invested his own time and credibility persuading them to sign on. Assad was summoned to an audience at the Borochov Ruchey summer residence prior to the summit and instructed to fall into line. Putin telephoned Trump to get the Americans on board. The American president was concerned only about counterterrorism and refugee return. That was exactly what Putin wanted to hear. Saudi Arabia's King Salman, Egypt's Fattah El-Sisi and Israeli's Benjamin Netanyahu were all brought on board by Putin personally. Only one major leader still objected, Turkey's Recep Tayyip Erdoğan: an autocrat like Putin, who saw the world in terms of power politics and zero-sum games. Although they often disagreed, Putin and Erdoğan spoke the same political language. Both understood and played for power. In return for supporting Sochi, Erdoğan wanted the Syrian Kurds excluded and a green light to take more territory from them, this time around Afrin in the northwest. Putin agreed.

Putin thus succeeded in getting 1,500 Syrians to Sochi and the major foreign players on board. But that proved his only success. Although he had the numbers, he did not have the *right* numbers. The principal opposition negotiating group refused to attend despite Turkey's urging. The armed groups which held most territory – the Kurds and jihadists – were excluded. Most of those that did attend were either loyalists or representatives of the so-called "loyal opposition" – neither capable of negotiating for the armed opposition. The few genuine opposition delegates to attend heckled Lavrov's opening address and refused to negotiate until government forces ceased firing on Idlib. They were cajoled into agreeing a constitutional committee but there was no agreement on its size, composition, or mandate. It was not the outcome Putin expected. His personal investment had yielded almost nothing. A new strategy was needed. "Plan C" involved winning Assad's war for him, the very thing Syria's president had wanted all along.

Plan C: Assad's plan A

The year 2018 would be the year Russia helped Assad win his civil war by taking the de-escalation zones one by one. At least, that was the plan. Turkey's President Erdoğan had other ideas. He understood how Putin operated better than any other foreign leader the Russian had yet faced (in that, he would be exceeded only by Volodymyr Zelensky). The military plan was to visit what had been done to Aleppo on the de-escalation zones: besiege, bombard indiscriminately until the area totally collapsed, take the area, filtrate the enemy – either by displacing them, imprisoning them, or killing them. Employed to devastating effect in Aleppo, it was a strategy of atrocity crimes strikingly like that employed in Chechnya nearly 20 years earlier.

The first to fall was eastern Ghouta on the outskirts of Damascus. In early February, intense artillery bombardment killed dozens of civilians. That was followed by a sustained artillery and air campaign. Russian aircraft joined the fray and the bombing continued without a break for three weeks, killing hundreds. Eastern Ghouta's defensive lines began to collapse in the first week of March 2018, splintering what was left of the enclave into three tiny pockets. Tens of thousands fled as the civilian death toll climbed above 1,500. The first of these pockets, Harasta capitulated around 22 March. A day or so later, the second capitulated when its frontlines collapsed. The defenders handed over control in return for the transfer of around 7,000 fighters and their families to Idlib. By the end of March, badly ravaged Douma was all that was left. The bombardment intensified in early April until, on 7 April, two chlorine bombs hit a bakery and a public square. Forty-three people were killed by exposure to the highly toxic chemical. The following day, the defenders caved in and accepted the government's terms. Those fighters that wanted to be were evacuated to Idlib. Between 12,000 and 18,000 people had been killed in the siege of eastern Ghouta.

Worried the chemical attack might provoke an American intervention, Russia orchestrated a macabre disinformation campaign. First, the Kremlin denied there had been a chemical attack. Then it claimed the White Helmets – a civilian rescue organization – had staged the attack. Then it blamed Britain, saying it had perpetrated the bombing. Then, in a grotesque homage to Stalin's 1930s show trials, it paraded 11 "witnesses" before the Organization for the Prohibition of Chemical Weapons (OPCW) at The Hague, each of whom claimed there had been no chemical attack. In fact, the OPCW found clear proof there had been a chemical weapons attack. The Russians needn't have worried, though, the American response was a cursory missile strike that did so little damage to Assad's war machine it was essentially a licence to proceed.

Proceed they did. The second de-escalation zone to fall was Rastan, the smallest and least defensible of the four. On 15 April, government forces units, backed by Russian air power, advanced into the zone. The beleaguered defenders quickly succumbed, negotiating a deal with the Russians to transfer fighters to Idlib. Around 3,000 fighters – with their light weapons, families, and other civilians – were bussed north to Idlib.

The third was the southern zone in Daraa, birthplace of the revolution in 2011. The southern de-escalation zone ought to have been far more difficult to overrun. Situated close to the Golan Heights, Israel watched the area carefully, opposed to any incursion of Iranian-backed forces. The southern zone had been guaranteed also by Jordan and the US. With a population of around 750,000, the southern zone was also far more populous than those already eliminated. An attack here risked serious escalation. In the event, Moscow skilfully prepared the way with Tel Aviv, the US keeled over, and the whole business was wrapped up before UN officials broke for their summer holidays. The offensive began on 18 June and government forces made rapid gains. Backed by Russian aircraft, they seized several towns and villages, displacing 50,000 from Daraa in the first week alone. Opposition forces launched counterattacks but failed to reverse the tide as, now confident neither the US nor Israel would intervene, Russia stepped up its fire support. By 5 July, approximately 60 per cent of the pocket was in government hands and around 160,000 people displaced. Close to 400 indiscriminate barrel bombs, (literally bombs built in barrels and tossed from planes and helicopters), 250 rockets, and a similar number of artillery shells had fallen on the zone since the offensive began. Civilian casualties mounted and the area's four main hospitals were deliberately attacked. A surrender deal like the previous two was agreed. Now, only Idlib remained.

Idlib was different. Since the vast bulk of the remaining armed opposition was concentrated there and had nowhere else to go, it would offer much fiercer resistance than the other three zones. More than 3.5 million civilians now called Idlib home. Then there was Turkey, its holdings around Afrin and further east directly abutting Idlib, and its allied militia dug in within and around the enclave. Above all, Erdoğan wanted stability in Idlib. Turkey could not afford for the enclave to collapse since that could trigger a new refugee crisis and present the Kurds an opportunity to grab more territory.

The first test was not slow in coming. The first offensive was launched in September 2018 with a surge of artillery fire and airstrikes – dozens conducted by Russian aircraft. Erdoğan hit back strongly; warning Turkey would not hesitate to defend the zone. As if to prove the point, Turkish troops and heavy weapons were deployed into the enclave. Erdoğan's obstinance and none too subtle threats persuaded Putin to cut a deal. Russia's president had always intended a limited offensive and was not prepared – yet – to risk escalation with Turkey. They agreed a ceasefire and demilitarized zone (DMZ) between government and rebel frontlines.

The Syrians and Russians tried again in the spring of 2019 when, in the final days of April, aircraft and artillery opened a barrage into the DMZ and wider enclave. Their principal goals were to clear rebels from the DMZ and secure positions behind some of Turkey's observation posts. Civilian infrastructure was deliberately targeted, especially hospitals. Eighteen hospitals were hit, four by Russian jets in a single day. Amongst the casualties were more than a dozen children, killed by barrel bombs that tore into schools, medical facilities and homes. After a week or so of bombing, Syrian government forces advanced into the DMZ. Turkey sent its Syrian allies into Idlib to bolster the defence, adding around 1,000 fighters equipped with T-72 tanks, TOW launchers, GRAD rocket launchers and anti-tank weapons. Meanwhile, at least one Turkish military convoy of more than a dozen vehicles supplied opposition fighters with TOW launchers and other critical equipment entered Idlib.

The civilian toll was immense. Between May and August some 400,000 civilians were displaced as government forces deliberately shelled civilian infrastructure inside the DMZ to terrorize the people there. Civilians displaced by the shelling found little safety elsewhere. On 26 August, fighter jets – most likely Russian – targeted a displaced persons camp, killing at least 20, including eight women and six children. The strike also damaged the camp's food store, medical centre and school. A string of attacks on medical facilities showed they were being targeted systematically.

In fact, the turning point had come a week earlier when government forces reached the outskirts of the prized town of Khan Shaykhun. Still hoping to avoid direct confrontation, Turkey tried to deter the government by despatching a 28-vehicle military convoy including tanks and armoured personnel carriers ostensibly to resupply two observation posts in the far south of the enclave. Ankara calculated that should the convoy reach Khan Shaykhun before the town fell, Damascus and Moscow would be deterred from proceeding. Watching the approaching convoy with alarm, Syrian government commanders took a massive gamble. Government aircraft launched two separate attacks on the convoy, killing at least one Syrian fighter allied to the Turks and three civilians, as well as injuring some Turkish soldiers. The attacks were plainly designed to delay Turkish reinforcement rather than inflict serious damage, bombs landing on the road ahead and to the side of the convoy rather than directly on it, so as not to inflict casualties and thus provoke Turkish retaliation. It worked. The convoy was halted and opposition forces withdrew from Khan Shaykhun. The sides agreed a new ceasefire.

If anything, Turkey's military position in Idlib was stronger at the end of the offensive than it had been at the beginning. Whether this meant that Erdoğan was now resolved to defend Idlib militarily, if necessary, remained unclear. That uncertainty troubled Putin. In fact, Turkey's president was indeed rethinking his Syria strategy. Turkey deployed additional forces so that if deterrence failed, Turkey could employ graduated escalations of force to impose spiralling costs on Damascus. Turkey's new strategy hinged on an assumption that direct confrontation with Russia could be avoided if it targeted only Syrian government forces. That involved a dangerous game of militarized chicken, since Russian officers largely controlled the Syrian army's active elite units, its land and air forces worked in close coordination with the Syrians, and Russian Wagner mercenaries (see Chapter 7) were embedded inside the Syrian army. By mid-February 2020, Turkey had between 7,000 and 20,000 soldiers and 2,000 armoured vehicles inside the Idlib enclave.

The final battle came in early 2020 with a remarkable display of brinkmanship between Erdoğan and Putin. Guided by Russian officers and intelligence, and supported by Russian airpower, government forces made steady progress which gradually turned into a rout. Turkish forces first attempted to deter escalation through presence. Ankara warned Damascus to back down and threatened a military response if it did not. Yet since both Assad and Putin still assumed it would be Erdoğan backing down, neither thought the threat serious. On 3 February, Syrian and Turkish forces exchanged fire. According to some reports, eight Turkish and double that number of Syrians were killed. Erdoğan declared the attack on Turkish troops a "turning point" and demanded Syrian withdrawal, warning of escalation if they did not.

Was Erdoğan's threat real or mere bluster? On 8 February 2020, a Russian delegation charged with finding out arrived in Ankara. They weren't convinced Erdoğan was serious, so the rout of Idlib continued. The town of Saraqib fell as opposition coordination collapsed. Victory seemed to be within Assad's grasp. The Russians and Syrians assumed that with a little more probing, Turkish resolve would collapse. On 10 February, government forces conducted a sustained mortar attack on a Turkish observation post at Taftanaz airbase. Turkish forces responded with a sustained barrage, which they claimed struck more than one hundred government positions, destroyed three tanks as well as other vehicles, and inflicted more than one hundred casualties. This was the very kind of step-up called for by Turkey's graduated strategy. But it did not stop the offensive. By the end of the week, government forces were celebrating significant advances and Assad gave a rare television address promising outright victory.

Now it was Russia's turn to increase the military pressure. A Russian-built elite Syrian unit, the Tiger Force, attacked Turkish positions on 23 February, backed by intense Russian and Syrian airstrikes. Turkish casualties climbed to ten and a Turkish drone was downed. Two more Turkish soldiers were killed the following day. But far from backing down as the Kremlin expected, Turkish artillery and drone strikes helped opposition fighters conduct an effective counter-offensive, causing shock and alarm in Moscow. Turkish forces were proving more than a match for the Syrian government. Still Putin assumed Erdoğan's resolve was weaker than his. It was a serious miscalculation.

On 27 February, surface-to-air missiles targeted Russian Su-34s. Syrian Su-22s and Russian Su-34s responded by attacking a military convoy, killing at least 33 Turkish soldiers. Erdoğan called Putin, angrily demanding an explanation. Putin denied Russian involvement and warned Erdoğan to accept his terms. Erdoğan refused. Shaken by that, Putin backtracked and proposed they agree to not allow differences on Syria colour their broader relationship. Erdoğan accepted the proposition and in public Ankara accepted that Russia had not been involved in the attack on its convoy. The two leaders also agreed to meet on 5 March setting the tenor and the timeframe for what was to come.

Turkish forces unleashed a barrage of artillery and drone strikes against more than 200 Syrian government and Hezbollah targets, including Russian mercenaries and proxies, causing hundreds of casualties and destroying dozens of tanks, APCs, artillery pieces, trucks and ammunition stores. Turkish missiles hit chemical weapons research facilities and at least two government-held airbases in Aleppo governate. Man-portable air-defence systems (MANPADS) targeted Syrian and (according to Russian sources) Russian aircraft with greater ferocity. In the skies, Turkish jets challenged Russian and Syrian government dominance and imposed a de facto no-fly zone over Idlib city. Turkish F-16s shot down two Syrian Su-24s, missiles and artillery rendered the government's Nayrab airbase inoperable. Turkish Bayraktar drones destroyed Russian-made Pantsir S1 anti-aircraft systems. The apparent ease with which Turkish drones jammed and then neutralized Russia's most advanced systems caused Moscow acute embarrassment – and shock.

By the time Erdoğan boarded his Moscow-bound aircraft for the prearranged 5 March summit with Putin, Turkey was claiming to have inflicted more than 3,000 casualties on the Syrian government and its allies, and to have destroyed or disabled three jets, eight helicopters, three drones, 151 tanks, 47 howitzers, 52 launchers, and more than 140 other military vehicles. These figures are disputed but are not wildly different to independent estimates. Turkey's intervention inflicted a heavy toll, stalled the government's offensive and exposed its military fragility. Turkish losses meanwhile were light, only a handful of soldiers and three drones.

There was no disguising the fact that Assad's position in Idlib – and with it Russia's – had been seriously weakened. Putin now faced the unenviable choice he had tried to impose on Erdoğan. Continuing its air campaign over Idlib would mean replacing lost Syrian capabilities with Russian assets or transferring more assets to the Syrians. It would also mean accepting greater losses for uncertain gains, something Putin – who had already achieved his principal objectives (saving Assad and reclaiming great power status) was reluctant to do. The two leaders agreed a new ceasefire that offered both a little of what they wanted but which represented a significant short-term victory for Erdoğan since it bought Turkey the time it wanted to shore-up Idlib's defences and political stability.

The lengthening arm

In the final analysis, Putin succeeded in reasserting Russia's place on the global stage by intervening in Syria. Coming a year after the annexation of Crimea, war in Syria supported the narrative of Russia's re-emergence as a peer competitor to the West. The appearance of victory reinforced confidence in the military and the Russian military's self-assurance. It also affirmed the president's belief in the utility of force and the weakness of a divided West. But in truth, it was a partial victory at best. Assad had exerted as much influence on Putin as Putin on Assad. The Kremlin found itself propping up a regime it knows cannot command the loyalty of most Syrians; a collapsed state run like a loose network of mafia fiefdoms facing a huge reconstruction bill it will be unable to foot alone. Putin accomplished neither his original goal, a peace deal on Russian terms, nor his revised objective of outright victory. Moscow's strategy had failed to contain jihadism and Moscow depended on the US-led coalition to defeat IS. In reality, a partial victory had been won over a badly fragmented rag-tag opposition which had nevertheless managed to resist for five years the combined might of Russia, the Syrian government, Iran, Hezbollah, and thousands of mercenaries.

Russia's geopolitical victory – its reclaimed status – also came at a cost: the alienation of a rival harbouring its own imperial pretensions: Turkey. The two powers came perilously close to war over Idlib and for all the Russian military's posturing the reality was that it had been consistently outperformed by the Turkish military. But victory is victory, and victory in Syria reinforced Moscow's growing hubris. Just a few months after the guns died down in Idlib, Russia and Turkey found themselves standing behind two different sets of combatants, this time in Nagorno-Karabakh.

Further reading

Rania Abouzeid, *No Turning Back: Life, Loss and Hope in Wartime Syria*. New York: Norton, 2019.

Alex J. Bellamy, *Syria Betrayed: War, Atrocities, and the Failure of International Diplomacy*. New York: Columbia University Press, 2022.

Sam Dagher, *Assad or We Burn the Country: How One Family's Lust for Power Destroyed Syria*. New York: Little, Brown, 2020.

Ohannes Geukjian, *The Russian Military Intervention in Syria*. Montreal: McGill-Queen's University Press, 2022.

Alexey Vasiliev, *Russia's Middle East Policy: From Lenin to Putin*. Abingdon: Routledge, 2020.

6
Nagorno-Karabakh

The 2020 war between Azerbaijan and Armenia over Nagorno-Karabakh is an outlier to the story told in this book. Vladimir Putin was neither the engineer nor instigator of war. Russia was not even a party. The 44-day war that erupted in late September was caused by a long-standing dispute between Azerbaijan and Armenia over who should govern Nagorno-Karabakh and the seven Azerbaijani districts occupied by Armenians since 1992. The war's wider, imperial, dimension was secondary to this primary, local, dimension. Yet that imperial dimension is an important part of our story.

Russia used the rivalry between Armenia and Azerbaijan to cement and attempt to extend its influence in the southern Caucuses. Although sentimentally predisposed towards Armenia, Russia attempted to maintain equidistance between the two rivals. This required a delicate balance since the geopolitical sands shifted significantly over time. Dependence on Russian security guarantees made Armenia a willing and loyal member of Russia's sphere of privileged interest, of the CIS, CSTO and Eurasian Union. Resource-rich Azerbaijan, meanwhile, wriggled free from Russia's orbit but not in a westward direction. Resource wealth and strategic location afforded Azerbaijan freedom to manoeuvre itself. Moscow needed Azerbaijan as much as Azerbaijan needed Moscow, forcing the Kremlin to deal with Baku on more equal terms. But Azerbaijan's increasingly authoritarian government had little inclination to embrace the West as anything more than a trade and investment partner. Autocracy thus placed normative distance between Azerbaijan and the West. To maintain influence amidst these changing conditions, Russian policy adapted to circumstance, sometimes at the cost of contradiction. It offered itself as peace mediator and seems genuinely to have pursued a settlement that could have resolved the conflict and established more stable conditions for the exercise of influence. Simultaneously, however, it sold weapons to both sides – an unusual stance for a mediator.

Amidst the delicate shifts and contradictions of policy, was one striking continuity. Among the Kremlin's enduring ambitions was the establishment of a Russian military presence inside Azerbaijan. Russia had just such a presence in Armenia. Fearful of both Azerbaijan and Turkey – memories of the 1915 genocide of more than a million Armenians during the death-throes of the Ottoman Empire understandably loomed large in the new Armenian state's political imagination – independent Armenia looked to Russia for security from the first. With the consent of Armenia's new government, the Soviet base at Gyumri, close to the Turkish border, became a Russian base, housing the 102nd Brigade. By this, Armenia got some teeth to back up the Russian security guarantees it gained by membership of the CSTO whereas Russia gained a means of projecting power. Things were different in Azerbaijan. The new state there expelled the Soviet army, leaving Russia with no military presence. During the First Nagorno-Karabakh War, Russia pressed both sides to agree to the deployment of Russian peacekeepers – a ploy used to install Russian forces in Georgia and Moldova too. Armenia agreed, Azerbaijan did not. With the return of war in 2020, Russia sensed a new opportunity to insert itself and this time succeeded.

Another imperial dimension was Russia's rivalry with Turkey. Armenia, formally allied to Russia through the CSTO, was challenged by an Azerbaijan allied to Turkey, another power harbouring imperial ambitions. In a sense, the 2020 war over Nagorno-Karabakh was the third round of a militarized contest between two imperial powers vying for influence across a wide geographic arc stretching from Azerbaijan to Libya. Erdoğan, whose Islamic leaning Justice and Development Party (AKP), came to power in Turkey in 2002, had overseen a period of unprecedented growth that had transformed the country's economic landscape and sense of itself. Emboldened by a growing economy and chastened by the EU's continued refusal to admit Turkey, Erdoğan's early Europeanism gave way to a very different vision of his country's place in the world, a vision described as "neo-Ottoman" by some commentators. This vision emphasized Turkey's unique position as a bridge between west and east capable of projecting influence in both directions. A Sunni Muslim majority society with a democratic government, secular state and thriving market economy, Turkey offered itself as a model for others to follow. Erdoğan's neo-Ottomanism was in many respects therefore a mirror-image to Putin's imperial vision. The two imperialisms clashed in Nagorno-Karabakh.

The Nagorno-Karabakh War was a primarily local affair between Azerbaijan and Armenia, but a war with imperial dimensions too. The way it was managed tells us something important about the transformation of European order. The First Nagorno-Karabkah War (1991–94) was mediated by the OSCE – the institutional form given to the "Helsinki" vision of European security. That mediation was an archetype of how disputes might be managed in the new Europe. The ending of the second war (2020) was quite different. It was mediated by Russia; Azerbaijan brought to the table by Turkey, Armenia forced to accept what it must. This was peacemaking redolent of "Yalta", not "Helsinki".

First Nagorno-Karabakh War

To understand what happened in 2020, we need to rewind 30 years to the ending of the Soviet Union and the first war for Nagorno-Karabakh. In Chapter 1, we left the region in an uneasy situation. The Soviet republic of Azerbaijan had deferred responsibility for quelling the rebellious Armenian controlled autonomous region to Soviet security forces. In April and May 1991, Red Army, Soviet Interior Ministry, and a hastily assembled Azerbaijani force (OMON – "Special Purpose Unit") had conducted Operation Koltso, a crackdown that entailed the encirclement of Armenian villages and forcible checks to root out members of the *fedayeen*, the volunteer armed groups that had sprung up to defend the secessionist cause. Smaller in scale and less systematically brutal than the filtration methods employed in Chechnya, the operation nevertheless followed a similar logic and there were numerous incidents of killing, torture, kidnapping, rape and forced deportation. Although it failed to seriously degrade support for the *fedayeen*, the operation did succeed in dampening the armed movement. But stability had come at a price. To persuade the Soviet military to move against Armenian separatists, Azerbaijan's communist leader, Ayaz Mutalibov, had aligned himself with the military's hardliners. Since Azerbaijan was a socialist republic with all the trappings of statehood but an uncertain connection to a national people and history, Mutalibov was prepared to sacrifice national independence to preserve the state's territorial integrity.

Armenia's nationalist leaders faced the opposite problem. They had a very clear sense of nation but a more uncertain sense of statehood. Many of Armenia's most precious historical sites lay beyond the Armenian republic's border, many to the west in what is now Turkey and from where Armenians had been evicted by bloody genocide but some of the most precious lay to the east, in Nagorno-Karabakh. Armenia's first post-independence leader, Levon Ter-Petrosyan, rose to prominence leading the campaign to unite Armenia with his native Karabakh. His successor, Robert Kocharyan was also a Karabakh Armenian. Thus was the movement for Nagorno-Karabakh's secession intimately related to Armenia's own national independence. Moreover, Armenian political affinities lay with Russian nationalists, that is, with Boris Yeltsin – the great rival to the Soviet hardliners backing Azerbaijan. The fate of remote Nagorno-Karabakh was thus intricately linked to Moscow's political drama.

Map 6.1 The Caucases, showing Nagorno-Karabakh
Source: iStock/PeterHermesFurian.

The attempted August 1991 coup disrupted the intricate political balance in the southern Caucuses and ignited war. When the coup collapsed Mutalibov lost the bulk of his military overnight while Armenia's nationalists meanwhile gained a powerful friend in Yeltsin. Azerbaijan's leader tried to shore up his rule by getting in ahead of the curve and in just two weeks, he declared Azerbaijan independent, was elected president, and dissolved its communist party – all without any change to the actual people who held power. Armenia followed closely behind. By the middle of October 1991, it too was independent and Ter-Petrosyan was president, head of a government dominated by Karabakh Armenians. Yeltsin tried mediating a deal between the two to forestall the seemingly inevitable but he lacked the means to make it stick and had problems of his own to attend to in Moscow. In late November, Armenian fighters shot down an Azerbaijani military helicopter, killing more than 20. A few days later, Azerbaijan's national council revoked Nagorno-Karabakh's autonomy. The first war had begun.

It was a war of three phases won decisively by Armenia. In its first phase, the war was irregular, chaotic and disorganized. Neither side had properly organized armed forces and instead deployed agglomerations of volunteers, interior ministry, police officers and mafia-like set-ups. The Armenians were the better organized because the *fedayeen* had been arranging themselves for years whereas the Azerbaijanis had until just a few weeks before relied on Soviet military power. Fighting was village to village rather than along frontlines, violence against civilians common since both sides aimed at "ethnic cleansing". What had once been an interlaced geography of Armenian and Azerbaijani villages was violently reorganized. The relative balance of forces differed from place to place but in this phase's principal operational action the Azerbaijanis used their position on the strategic heights of Shusha to envelop and bombard the separatist capital just a few kilometres away at Stepanakert.

The siege was not effective and in early 1992, the Karabakh Armenians broke out, seizing a string of Azerbaijani villages, expelling the residents as they made for Khojaly and its precious airport. Khojaly had been the focus of earlier Azerbaijani attempts to change Nagorno-Karabakh's ethnic balance and in the preceding years more than 6,000 Azerbaijanis expelled from Armenia had been resettled there. As Azerbaijani civilians fled the Armenian advance, they were hit by a hail of mainly Armenian gunfire. It is alleged that elements of the Soviet

366th regiment, freelancing with the Armenians for pay and food, also participated in the bloodletting. Moscow certainly seemed to think so, ordering the regiment's immediate withdrawal, and then disbanding it soon after. More than 200 civilians, and perhaps more than 400, were massacred. Even the lower estimates mark out Khojaly as by far the largest massacre of the war. It caused a political storm in Baku. Mutalibov, accused of failing to protect his people, was forced to resign. Azerbaijanis inflicted their own revenge inside Nagorno-Karabkah. Around 50 Armenian civilians were massacred in Maraga a few weeks later.

Not for the last time, the Armenians exploited the chaos in their enemy's ranks. Strategic Shusha, the symbolic heart of Azerbaijani Nargorno-Karabakh, fell to them quickly, lifting the siege on a Stepanakert devastated by shellfire. Shusha's defenders, Shamil Basayev among them, complained they had been abandoned by Baku. The Armenians also managed to connect their territory to Armenia proper by taking Lachin. By mid-May 1992, virtually all Nagorno-Karabakh lay in Armenian hands and virtually all its Azerbaijani residents had been forcibly displaced. Baku itself teetered on the brink of collapse as rival factions, backed by different elements of the security forces, vied to either restore or replace Mutalibov. Order of sorts was restored in June with the election of nationalist Abulfaz Elchibey to the presidency.

The war's second phase opened with Azerbaijan resurgent. Seemingly united behind their new president, Azerbaijani forces retook almost half the disputed territory by September. The apparent shift in the military balance was accompanied by two other changes. The use of armoured vehicles and tanks to spearhead Azerbaijan's offensive indicated that the war was developing into a conventional interstate war. This was enabled by the new Russian authorities in Moscow agreeing to turn over vast quantities of Soviet arms to the two new states, relieving itself of the burden of repatriating or maintaining an arsenal it could no longer afford. Azerbaijan learned quickest how to use the equipment placed at its disposal but it was not long before the Armenians caught up. In a further indication of just how chaotic these times were, the tanks which led Azerbaijan's line were crewed by Russians, specifically members of the 4th Soviet Army based in Ganje, not Azerbaijanis. Abandoned by a government dissolved, unpaid and sometimes even unfed, many Russian soldiers (and some Ukrainians and Belarussians too) were happy to find employment and income with the Azerbaijanis. Meanwhile, other

Russians (and Ukrainians and Belarussians) found employment with the Armenians. Thus, in the twilight between the endtimes of the Soviet Union and beginning times of the new states that followed, Russians fought Russians on behalf of Azerbaijanis and Armenians in the mountains of Nagorno-Karabakh.

The great Soviet army carve-up initially favoured Azerbaijan since more Soviet troops had been stationed there than in Armenia. Moscow tried to balance that out by selling arms to Armenia at a discount. The Azerbaijanis complained about Russian partiality. Certainly, Yeltsin sympathized with the Armenian cause and probably hoped to achieve a military balance leading to a stalemate and political deal, but whether Russian support extended beyond that remains unclear. The Kremlin wanted to keep both Armenia and Azerbaijan within its sphere of influence as members of the CIS and wanted a peace deal that would legitimize a continuing Russian military presence. But Azerbaijani nationalist objections about what they saw as Moscow's partiality towards Armenia pushed both objectives beyond reach. Elchibey rejected CIS membership and rebuffed any suggestion of Russian peacekeepers. He had reason to be confident as by the end of 1992 the war was going his way and it was Armenia heading towards economic collapse.

The war's second phase ended with Azerbaijan ascendant. But it did not stay that way for long for reasons that remain unclear. Armenia did not buckle under stress. Where Baku under pressure had fallen into factional fighting, no such thing occurred in Yerevan. Armenians may have had no money, but they did have a strong sense of shared identity and purpose still embodied by Ter-Petrosyan. Russian-supplied arms also helped and Armenian forces learned to make better use of the arms they had. The fiction that the war was internal to Azerbaijan – a war between the government and the Karabakh Armenians – was quietly abandoned as Armenia's new army, fortified by volunteers from Armenia and the diaspora, became fully engaged. But the most significant factor of all was internal dissent within Azerbaijan and a struggle for power that involved the Kremlin but in a way that to this day remains opaque.

There were signs in early 1993 that one of Azerbaijan's top commanders in Nagorno-Karabakh, Surat Huseynov, was following his own path not that of the government. Areas were left undefended, embattled units not reinforced. Rumours suggested Huseynov and defence minister Rahim Grazyev were mobilizing against the president and that both were working with the Russians. Huseynov's unruliness weakened Azerbaijan's grip on Karabakh. His ambition seems to have been to unseat Elchibey but in favour of who, and why? No one knows for sure, but one plausible theory suggests Huseynov and Grazyev were working with the Russians to engineer Mutalibov's return, a leader the Kremlin thought it could rely on to end the war, invite Russian peacekeepers, and bring Azerbaijan into the CIS. Tensions escalated in February 1993, when the loss of two key villages provoked accusations about Huseynov's loyalty. The commander pulled his forces from the front, weakening Azerbaijan's position. Elchibey ordered the units dissolved. Grazyev was sacked. But Huseynov resisted Baku's orders, led his troops back to base in Ganje, and waited. In June, Elchibey sent troops commanded by new defence minister, Dadash Ryazev, to disarm Huseynov's rebels. But instead, Huseynov's battle-hardened troops disarmed the relatively inexperienced government forces and headed for the capital, seemingly intent on unseating the president. Now, Elchibey did something no one had anticipated, flying to Nakhchivan, Azerbaijan's exclave bounded on three sides by Armenia, the homeland and personal fiefdom of Heydar Aliyev. Aliyev loomed large over Azerbaijani politics. For 18 years he had served as party first secretary and effective leader of Soviet Azerbaijan until he had been deposed by Gorbachev. Aliyev was Azerbaijan's political giant; out of government but never idle. Elchibey hoped Aliyev could keep his rivals out of power and the former leader did just that. His return to Baku undercut Huseynov and Mutalibov. Aliyev was elected president in October, with 77 per cent of the vote, a position he never relinquished. The wily president tried making peace with Huseynov by bringing him back into government as prime minister, but the renegade commander never fully reconciled with Baku. He attempted another coup in 1994 but was outmanoeuvred and imprisoned. He was later released into exile in Russia.

Although Aliyev was winning the war against state collapse, he was losing the war for Nagorno-Karabakh. In its third phase, the war became more conventional, more deadly, and much more one sided. The tide had started to turn whilst Huseynov was still at the front. When his troops withdrew, a trickle of setbacks turned into a flood. Martakert fell in June. Aghdam in July. The Armenians reached beyond Nagorno-Karabakh into Azerbaijan itself. Kelbajar fell in April. By October 1993 Armenians held the whole of Nagorno-Karabakh and seven regions around it.

Aliyev was not finished yet, however. The Azerbaijanis launched a huge winter offensive in the bitterly cold months from December to February 1993–94. Now a war between two evenly matched conventional forces arrayed with air power, artillery and missiles, these were the bloodiest months of the three-year-long war. An entire Armenian battalion (more than 300 soldiers) was surrounded and destroyed in January. The next month, an Azerbaijani offensive on Kelbajar was forced into a disorderly retreat across snow riven mountains. Hundreds were killed. Neither side could sustain these losses for long and although the violence was intense, more intense than anything that had gone before, relatively little ground changed hands. Armenians held on to the territory they had taken. By the spring, both sides were ready to negotiate.

The principal negotiating vehicle was the "Minsk Group" (whose membership eventually settled on Russia, US and France) established almost as an afterthought at the OSCE's inaugural conference in 1992. This formal OSCE process cohabited with a separate but related Russian mediation. Not until mid-1993 did the Russians have a coherent policy extending beyond the vague aspiration of maintaining its own regional supremacy, preferably by both countries joining the CIS, both regarding Russia as their chief interlocutor, and both accepting a Russian military presence on the ground. The Huseynov affair suggests Russia's defence ministry at least had not yet entirely given up on the possibility of establishing a presence in Azerbaijan beyond mere peacekeeping. The Kremlin thus favoured peace but this wasn't an exclusive priority. When it seemed Azerbaijan would prevail, Russia stepped up military support for Armenia and probably meddled in Baku's internal politics. But the Kremlin stopped short of overtly picking sides. It sold arms to Azerbaijan and refused to give Armenia the means to win outright, a difficult balance by which it hoped to maintain influence over both. Russians did most of the heavy lifting for the ceasefire negotiations through the first five months of 1994, and right until the last they advocated the deployment of Russian peacekeepers as part of the arrangement. Although Mutalibov may have relented on that, the more astute Aliyev would not. Whatever his own feelings, Aliyev knew he would lose the nationalist support he depended on if he conceded to Russian peacekeeping. Meanwhile, though they welcomed the idea of Russian peacekeepers, the Armenians felt them unnecessary. What mattered more was the buffer zone around Nagorno-Karabakh they now controlled. So it was

Russia that was forced to concede. A ceasefire without Russian peacekeepers was agreed on 12 May 1994.

Sands shift

There is debate about whether Russia kept the Nagorno-Karabakh conflict frozen because that served its own interests better. It is true that neither Yeltsin nor Putin expended much effort on the search for a more permanent resolution, but the notion that Russia actively sought to keep the conflict unresolved should be set against three other considerations. First, there were limits to what Russia could achieve, as demonstrated by its failure to persuade Azerbaijan to accept Russian peacekeepers. At no point between the two Nagorno-Karabakh wars was there a groundswell of support for conflict resolution in either Armenia or Azerbaijan if that meant making compromises. Although leaders in the two countries periodically moved towards an accord, as in 2001 when the US brokered talks in Key West and in 2009 when the "Madrid principles" were agreed, they faced significant domestic opposition whenever they did. Proposed arrangements fudged rather that addressed the central issue of Nagorno-Karabakh's political status.

Second, unlike in Georgia and Moldova, the Kremlin didn't want to leverage influence over one government but two. This complicated things immensely. Armenia's security dependence made it a reliable ally, but Moscow wanted equally good relations with Azerbaijan too. Strategically, Azerbaijan had always been more important than Armenia thanks to its location, size, wealth and resources. Oil and gas gave it political independence too. In 1997, Azerbaijan concluded a lucrative deal with a consortium of Western companies to exploit new fields under the Caspian. Meanwhile, it shed its dependency on Russian infrastructure by establishing pipelines to the west running through Georgia to the Black Sea and Turkey, avoiding Russia. Baku became central to plans for a trans-Caspian gas pipeline conveying gas from Turkmenistan and Azerbaijan to Europe again bypassing Russia. The 2000s' resource boom fuelled Azerbaijan's wealth just as it did Russia's, and the pipelines allowed it to forge ties with the West independently of Moscow. It also made Azerbaijan immune to Russian energy and economic coercion. Meanwhile, Aliyev wanted to rule Azerbaijan as Putin ruled Russia. That is, through a personalized autocracy. He did not want EU membership with its burdens of democracy, good governance and human rights and saw little need to join NATO. Before his death in 2003, Heydar Aliyev engineered the succession of his son, Ilham. His disregard for democracy and human rights only endeared him to Putin, who saw in the new Aliyev a stable autocrat he could do business with. The problem was that Aliyev junior wanted Nagorno-Karabakh back and the longer the problem went unresolved, the more likely it was that Azerbaijan would look elsewhere for help. If not Moscow, perhaps Ankara? Although Armenia could be relied on to stay inside Russia's sphere of influence, the conflict would have to be resolved if Russia wanted to keep Azerbaijan there too.

Third, despite all that, Russia did more than most others to mediate a solution. The Kremlin hoped to extend its influence by playing the role of mediator and peacekeeper and thus encourage Yerevan and Baku to look to Moscow not the West. A peace deal might yield agreement on Russian peacekeepers or encourage Azerbaijan into the CIS or – later – the Eurasian Union. At the very least, it would prevent a situation where Russian might be called upon to deliver on its CSTO security guarantee to Armenia in the event of war reigniting. Russian resource companies also favoured a peace deal that would enable them to expand their portfolios in the region. Putin himself reached out during his first 18 months in office but quickly lost interest when his efforts were rebuffed. Nevertheless, the Kremlin played a useful role in the failed Key West initiative after which George W. Bush lost interest too. The 2008 war in Georgia provided fresh impetus to the Kremlin's efforts. Having seemingly lost Georgia, it became suddenly more important to keep Azerbaijan and Armenia. President Medvedev held talks in Moscow with Aliyev and Armenian president, the nationalist Serzh Sargsyan, throwing himself into the process with unusual vigour but familiar results. Having failed himself, Medvedev passed the baton to Sergei Lavrov who fared no better, thanks largely to Armenia's commitment to the status quo. That Putin, Medvedev and Lavrov failed reinforces the first two points and introduces another: the conflict didn't rank very high on the Kremlin's list of priorities. Although Putin would have preferred a peace deal, he judged the risks of trying to unfreeze the conflict against the wishes of one of its parties outweighed the likely benefits to Moscow.

However, the tectonic plates beneath the conflict were shifting. Armenia won in 1994 because it had marshalled its military resources better and capitalized on Azerbaijan's political chaos. Azerbaijan achieved political stability and immense wealth in the 2000s, which it converted into military power. It also got a powerful ally in Turkey. The resources boom stimulated economic growth, hovering around 10 per cent for the first decade of the twenty-first century and between 2–5 per cent for the second. Besides an orgy of skyscraper construction that transformed Baku's skyline, the resource boom fuelled a massive increase in defence spending. Where the two sides had spent roughly the same on defence during the first Nagorno-Karabakh war, by 2020 Azerbaijan's defence spending was three and a half times greater than Armenia's. Armenia was spending close to 5 per cent of its GDP on defence yet still falling hugely behind. Azerbaijan invested especially heavily in capabilities designed to overcome Armenia's defensive advantages: unmanned aerial strike capabilities, long-range artillery, cyber and intelligence capabilities, and the command and control systems needed to bring this all to bear. Just how far Azerbaijan had come became apparent only in 2020.

If Azerbaijan's politics ran from chaos to authoritarian stability, Armenia's ran in the opposite direction. Government there became more autocratic but much more unstably so than in Baku. Serzh Sargsyan used electoral fraud and constitutional change to maintain power into a second decade, but he had no economic miracle with which to burnish his legitimacy, which continued to rest on nationalism and Armenia's unyielding claim to Nagorno-Karabakh. Armenia was becoming less powerful relative to Azerbaijan and more dependent on Russia. Almost all Armenia's weapons and ammunition came from Russia, much of it at discount rates. Armenia came to rely on the deterrence offered by Russia's security guarantee through membership of the CSTO and Russia's military presence at Gyumri. Yerevan paid a price for its security dependence, however, forced by Moscow to walk away from an association agreement with the EU and join the Eurasian Union instead, later accepting a much-truncated EU partnership.

By contrast, Azerbaijan was much more powerful relative to Armenia and much less dependent on Russia. It made a concerted effort to diversify its sources of arms, spending heavily on advanced unmanned aerial systems from Turkey and Israel. After 2015, only around one-third of the country's procurement budget was being spent on Russian weapons. Azerbaijan's main security partner now was Turkey. They shared a common faith (albeit Azerbaijan being mainly Shi'a and Turkey mainly Sunni) and mythic history that allowed them to conceive a common political project independent of Russian influence. As early as 1995 Aliyev senior had described them as "one people with two states". After Syria, Erdoğan sought to challenge Putin wherever he could. Azerbaijan and Turkey shared a mutual antipathy towards Armenia – Turkey's efforts to normalize relations with Armenia fizzled out over the genocide question in 2011. And they shared economic interests – Azerbaijan gave Turkey access to the Caspian and Central Asia; Turkey gave Azerbaijan access to the west.

These shifts in the balance between Armenia and Azerbaijan weakened the foundations of the 1994 freeze deal. With preponderant military power and a powerful ally, Aliyev spoke openly of using force to reclaim Nagorno-Karabakh. Nationalist rhetoric was popular at home and might also have been intended to threaten Armenia into compromising. Putin's position was consciously ambiguous. The Kremlin signalled opposition to the use of force but also clarified that its security guarantee applied only to Armenia proper and not the disputed territories inside Azerbaijan.

Azerbaijan tested the water in July 2016, when it attacked Armenian positions along several points of the frontline. The offensive quickly demonstrated Azerbaijan's new capabilities but after losing ground in the first two days, the well dug-in Armenians inflicted significant losses on the third, retaking most of the lost ground. On the fourth day, they agreed a ceasefire that awarded a small amount of land to Azerbaijan. Both sides lost around 100 soldiers killed and more than 500 wounded. Analysts disagreed about whether the approximately 15 sq km of land exchanged was strategically important. If Azerbaijan intended to browbeat Armenia by a demonstration of overwhelming force, it failed – Armenian defences held up. If its intent was to elevate global interest, that failed too – world attention proved fleeting. The skirmish did allow Aliyev to claim a victory – the first time in more than two decades that Azerbaijan had retaken territory. But it caused friction in Baku's relationship with Moscow. Privately Putin blamed Azerbaijan for the escalation and saw Turkish hands behind it. Lavrov publicly accused Turkey of fermenting war.

Although the Kremlin was clear that the CSTO's security guarantee applied to Armenia not Nagorno-Karabakh, there was likely sufficient ambiguity in its position to deter Azerbaijan and Turkey. The credibility of that deterrent, however, hinged on Armenia maintaining good relations with Putin. That relationship, however, was rocked by Armenia's Velvet Revolution.

As it became less democratic, so Sargsyan's government also became less popular. In April 2018, an announcement that Sargsyan intended holding onto power for a third term by switching from the presidency to the prime ministership (Putin-style) provoked mass demonstrations in Yerevan. Days of tumult brought opposition activist Nikol Pashinyan to power. Pashinyan was not an avowed westernizer but a former ally of Ter-Petrosyan. Armenia's new leader was careful to insist that his was not a colour revolution, something that went down well in Moscow but could not completely obscure the fact that another leader had been toppled by the people. Although he courted Putin, Pashinyan had always been more cautious towards Russia than his predecessor. That he had opposed Armenia's accession to the Eurasian Union in 2013 and raised the possibility of Armenia's exit from the union since did not go unnoticed in the Kremlin. As a member of parliament, Pashinyan had voted against deepening military ties with Moscow, and cast doubt on the credibility of Russian security guarantees. Pashinyan's rise may not have represented an Armenian tilt towards Europe, but it did raise questions about Yerevan's intuitive dependence on Moscow. Putin meanwhile had grown accustomed to Armenian loyalty, which the Velvet Revolution placed in question. It added to a combustible mix of shifts of power and allegiance and a conflict freeze thawed by violence in 2016.

Second Nagorno-Karabakh War

The year 2016 taught Azerbaijan that force could change facts on the ground in Nagorno-Karabakh without necessarily provoking Russian intervention but also that more work was needed to translate military assets into decisive capability. Those realizations drove Baku closer to Ankara. For Turkey, this relationship needs to be understood in the context of its struggles with Russia in Syria (Chapter 5) and Libya (Chapter 7). Only in 2020 did the full extent of the military ties between Azerbaijan and Turkey developed after 2016 become apparent. With Turkish help, the cultural and doctrinal orientation of Azerbaijan's general staff was shifted away from Russia. Officers were retrained, many sent to Turkey for re-education. Those judged too wedded to the old Russian ways or too connected to Moscow were purged. Turkish military intelligence was embedded inside Azerbaijan's military, giving the latter access to the eyes and sensors it would need to direct precision weapons. In Armenia meanwhile, Pashinyan advocated a new "non-ideological" politics beholden to neither East nor West. In early 2020, Putin tried to contain the Armenian's enthusiasm for non-ideological ideology by reminding him once again that the CSTO guarantee applied only to Armenia, and not Nagorno-Karabakh. Pashinyan also struggled to carve out a space for progress on Nargono-Karabakh. He came to power promising to resuscitate the peace process but that fizzled out after just three informal meetings with Aliyev thanks to domestic opposition and his decision to prioritize political reform. Pashinyan pivoted to nationalism to shore up his political position and pledged support to the goal of uniting Nagorno-Karabakh with Armenia. To Aliyev, that seemed to rule out any prospect of a negotiated resolution.

The situation escalated again in July 2020. Accounts differ as to what triggered the fighting. The Azerbaijanis say that the Armenians forcibly reclaimed an abandoned border post; the Armenians say the Azerbaijanis sent military vehicles too close to one of their checkpoints. It is likely both contain some truth. But whatever the cause, conflict erupted at several points along the Armenia/Azerbaijan border, including along Armenia's border with the Nakhichevan exclave. The two sides exchanged fire for four days. Both claimed victory although losses were relatively even, and no territory changed hands. Azerbaijan demonstrated its new unmanned aerial capabilities, destroying at least one Armenian tank; Armenia demonstrated the potency of its Russian air defences, downing 13 drones. Perhaps the most significant thing to happen, however, was the killing of Azerbaijani General Polad Hashimov. The public demanded revenge. Azerbaijan prepared to deliver it.

In July and August, Turkey and Azerbaijan conducted joint military exercises in preparation for an offensive which opened in the early hours of 27 September, just hours after the end of Russia's "Kavkaz 2020" military exercises in the northern Caucuses. Baku's objective was to reclaim as much territory as possible before Moscow and/or the EU imposed a new ceasefire. The timing in relation to "Kavkaz 2020" cannot have been coincidental and lends credence to the idea that Moscow was notified in advance for it seems inconceivable that Azerbaijan and Turkey would have acted without obtaining assurances of Russian non-intervention. That would also fit the pattern of Russian–Turkish behaviour in Syria where force and diplomacy continued side by side. We might also speculate that other terms were set in advance too – for example, measures to ensure the CSTO guarantee was not triggered, and that the offensive be limited to the seven lost Azerbaijani regions rather than the capture of Stepanakert since a complete Armenian reversal might trigger demands for intervention that Moscow would find hard to resist.

If Moscow did know what was coming, it didn't tell the Armenians. The campaign opened with an intense artillery and missile barrage that overpowered Armenian air defences and gave the Azerbaijanis air superiority. The ground offensive focused initially on the south and was spearheaded by around 1,000 Turkish-trained special forces. Although both Ankara and Baku deny it, there is strong evidence that Turkey recruited some 2,000 Syrian mercenaries to fight in Azerbaijan's spearhead too. Reports suggest they sustained heavy casualties, somewhere between 200 and 500 dead – a high attrition rate even at the lower estimate. Turkish officers assisted in planning the campaign, assessed and advised on operational readiness, and Turkish surveillance aircraft operating in Turkish airspace, Turkish satellites, and Turkish surveillance drones, monitored the conflict zone and fed real-time intelligence to the Azerbaijanis to guide their targeting. Six Turkish F-16s were also deployed into Azerbaijan. The Azerbaijanis made rapid progress, especially in the south capturing Jabrayil on 4 October and Hadrut, from which it was possible to control Nagorno-Karabakh's road link to Armenia, a few days later. By mid-October, it had retaken the full length of its border with Iran and looked set to advance on the Lachin corridor.

The improvement of Azerbaijan's military capability and the degree of Turkish involvement came as a surprise to most. Another surprise for Pashinyan and the Armenians was Russian ambivalence. Not only did Russia not furnish Armenia with active military support to counter the aid Azerbaijan received from Turkey, it also didn't resupply weapons and ammunition. Nor did it even offer full-throated political support. What it did do was propose a ceasefire that ratified Azerbaijan's territorial gains but even that was a half-hearted effort not backed by arm-pulling, incentive offering, or coercion. There were likely three reasons for that. Putin wanted to remind Pashinyan of Armenia's dependence on Russia, to cultivate his warming relationship with Aliyev, and allow events to play out that might create new opportunities for Russia to insert itself on the ground. An additional speculation is that Putin was already contemplating the invasion of Ukraine and thus wanted to avoid additional entanglements, and that the Russian military was in no position to support Armenia whilst simultaneously preparing for war with Ukraine. It is also likely that Putin was as surprised as everyone else at just how decisive Azerbaijan's early moves were and just how deep its military ties with Turkey had become. With its back against the wall, Armenia fired a handful of missiles, possibly including its advanced Russian-made Iskander missiles, into Azerbaijan, perhaps hoping that Baku's retaliation might force the CSTO into action. Azerbaijan did retaliate, using UAVs to target missile launchers inside Armenia but the Kremlin insisted that since this was part of a war over Nagorno-Karabakh the CSTO's guarantee was not triggered. After the initial exchanges, the missile war petered out.

The orientation of Azerbaijan's offensive shifted towards Shusha in early November. Quite why remains unclear but it would have been difficult to manoeuvre into Lachin without engaging forces stationed in Armenia itself and if successful would have choked Nagorno-Karabakh completely, the sort of escalations that might have forced Russia's hand. Azerbaijan was seemingly trying to achieve as much as possible without provoking Russian intervention – further suggesting that Baku had a reasonable idea about where the Kremlin's red lines were. However, destroying Armenia's air defences and taking low-lying areas abutting Nagorno-Karabkah was one thing, taking mountain-bound and well-defended Shusha another thing entirely. The Armenians inflicted heavy losses on Azerbaijani forces manoeuvring for territory and a frontal assault on Shusha was repelled. Meanwhile, now fearing the complete collapse of Armenian positions, Russia increased military aid to nullify some of Azerbaijan's advantages. Armenia was given Polye-21 electronic warfare capabilities and advanced mobile air defence systems (including Buk-M2), while Russian forces in Gyumri jammed Azerbaijani reconnaissance drones and signals. The result was a sharp increase in Azerbaijani drone losses and commensurate decrease in their capacity to fix and destroy Armenian defences.

Azerbaijan turned to more traditional methods. Azerbaijani special forces walked for five days using night cover and fog to infiltrate Shusha by climbing sheer rock faces at undefended points. Once inside, they quietly regrouped and then attacked at multiple points from behind Armenian positions. The confused and panicked Armenians fell into disarray and Shusha quickly fell, a critical moment not just for its strategic importance – Shusha being a mountain fortress in the heart of Nagorno-Karabakh looking down upon Stepanakert – but also for what it symbolized, for Shusha was the emotional heart of Azerbaijani Nagorno-Karabkah. Its fall triggered a flurry of Russian-led diplomacy and this time the defeated Armenians were ready to talk.

The deal engineered by Putin on 9–10 November confirmed a decisive victory for Azerbaijan. It called for a ceasefire, the withdrawal of all Armenian regular forces from Nagorno-Karabakh, the ratification of Azerbaijan's territorial gains and transfer to Azerbaijan of all territory in the seven regions surrounding Nagorno-Karabakh still in Armenian hands (leaving the Armenians with about 70 per cent of Nagorno-Karabakh and nothing besides the Russian-patrolled Lachin corridor beyond), the reopening of road links between Azerbaijan and the Nakhchivan exclave, and the deployment of 2,000 Russian peacekeepers for five years to monitor the ceasefire and the Lachin corridor, which the Armenians would hand over to the Russians. This insertion of Russian peacekeepers was a longstanding Russian objective, yet peacekeeping in Nagorno-Karabakh affords Russia few military advantages. Its peacekeepers are isolated and dependent on Azerbaijan, since there is no border with Russia and Russia's base in Armenia is some distance away. Whatever influence the peacekeepers give Moscow over Baku is therefore balanced by the influence running in the other direction. The immediate political effect of the peacekeeping deployment was to transfer responsibility for Nagorno-Karabakh's security from Armenia to Russia. As if in recognition, the self-styled Armenian Republic of Artsakh there elevated Russian to the status of an official language. For Russia, that is simultaneously a source of leverage and a burden.

Since Azerbaijan had won a more decisive victory than anticipated, it was able to impose additional limits on Russian peacekeeping. The mission was limited to a fixed five-year term, after which Baku's consent would be needed. Moreover, Russian peacekeepers would be accompanied by a small group of Turkish military observers. Although numbering less than 50, this represented a major geopolitical shift since not only was Turkey now exercising military influence directly in the southern Caucuses, but Putin had been forced to acknowledge that fact. Russia's president expressed his irritation outwardly, blaming the Armenians for not accepting his original ceasefire plan. He clearly felt that Russia could have got a better deal then, too.

Consequences

Azerbaijan emerged victorious from the Second Nagorno-Karabakh War. Its military offensive went further and faster than Putin (or anyone else outside Azerbaijan and Turkey) anticipated and thus squeezed out a better deal for itself. It seems likely that the war was in some sense managed by Russia and Turkey according to informally understood principles. That, after all, was how they had managed affairs a few months earlier in Syria: military competition to ascertain relative power and kept within certain political boundaries. If we accept this, Azerbaijan's victory seems even more complete. It succeeded in reclaiming the territory it set out to reclaim, ejected the Armenian army, and put limits on Russia's regional pre-eminence by helping Turkey, the other big winner, to establish itself more firmly. Just months after Idlib, Turkey scored a second victory in its great imperial game with Russia. Erdoğan joined an ebullient Aliyev for Baku's victory parade and recalled Aliyev senior two decades earlier in describing their relationship as a "brotherhood". Pashinyan's government, meanwhile, was thrown into immediate crisis by the defeat and the prime minister resigned in the face of mass protests before staging a remarkable comeback by winning the June 2021 elections.

Most observers thought Putin a winner too. Russia achieved its goal of deploying peacekeepers, albeit with strings attached; had prevented Azerbaijan from taking Stepanakert and causing a humanitarian catastrophe by displacing all of the 130,000 or so Armenians living in Nagorno-Karabakh; and had done so without ruining relations with Aliyev, Turkey, or Iran. Putin also delivered a broadside to Armenia's velvet reformers, reminding them of their dependence on Russia though this too proved a more ambivalent message than Putin would have liked since its reluctance to aid Armenia cast doubt on the

CSTO's credibility. Putin succeeded in cutting the West out of the deal, although he was unable to position Russia as sole arbiter and was forced to admit Turkey as a partner. Still, that represented a shift from the primacy of the OSCE's "Helsinki" model of security management towards a form of great power managerialism redolent of Yalta. But although a Russian–Turkish condominium in the southern Caucuses sits more easily with Putin than the institutional pluralism of Helsinki, it is a distinctly second-best outcome to the dream of Russian primacy in its "zone of privileged interests".

The Second Nagorno-Karabakh War was a war between Azerbaijanis and Armenians in which the imperial dimension was secondary. Within that imperial space, Putin's vision of great power Russia resurrected crashed into another's imperial dreams, that of Recep Tayyip Erdoğan's neo-Ottomanism. As in Syria, Putin and Erdoğan competed within boundaries they both understood. As they clashed, the vision of world order the two leaders shared, premised on the values of Yalta, displaced that of the OSCE's Helsinki vision of a rule-bound institutionalism. For them, war was a mere continuation of their politics, but by other means, as Prussian strategist Carl von Clausewitz put it in the nineteenth century. War and politics thus merged into one another: war an extension of politics; politics an extension of war. Putin has always seen politics this way, without neat distinctions between war and peace, force and diplomacy. It's a game Erdoğan understands well. The West took longer to comprehend Putin's logic, to understand that Putin's Russia was already at war.

Further reading

Laurence Broers, *Armenia and Azerbaijan: Anatomy of a Rivalry*. Edinburgh: Edinburgh University Press, 2019.

Charles King, *The Ghost of Freedom: A History of the Caucuses*. Oxford: Oxford University Press, 2008.

Jeffrey Mankoff, *Empires of Eurasia: How Imperial Legacies Shape International Security*. New Haven, CT: Yale University Press, 2022.

Anna Ohanyan and Laurence Broers (eds), *Armenia's Velvet Revolution: Authoritarian Decline and Civil Resistance in a Multipolar World*. London: I. B. Tauris, 2020.

Thomas de Waal, *Black Garden: Armenia and Azerbaijan Through Peace and War*. New York: New York University Press, 2013.

7
Shadows

The West may not have known it, but Putin's Russia was at war with it long before missiles rained down on Kyiv in 2022. It was not a conventional style of war. After the annexation of Crimea, Western analysts had scrambled to put a label on what seemed to be a new Russian way of war. "Hybrid war", "grey-area war", "full-spectrum war", "asymmetric war", the "Gerasimov doctrine" were just some of the terms devised. All these labels captured some of the truth, but none told the whole story. One way of thinking about the war ethos that gripped Putin's Kremlin after he returned to the presidency in 2012 is offered by seventeenth-century English political theorist, Thomas Hobbes. Hobbes described a condition of anarchy, which he called the "state of nature", where people were forced to compete in a zero-sum struggle for survival. The result was a permanent war of all against all. This was not a state of permanent battle but rather a disposition. Each person distrusted every other; each believed the other posed a threat and presumed that increments of security could be bought only at the expense of others; and thus force was a latent possibility in every social interaction. What mattered most for Hobbes was this disposition, not the instruments used to conduct the war.

This image well describes how third-term Putin understood Russia's relationship with the West. He assumed the West was waging hybrid war against Russia, using its soft power, economic levers, clandestine force, and sometimes its military might to forcibly change regimes and extend its hegemonic reach. Since NATO could not risk direct war with Russia owing to its nuclear arsenal, it seemed to the Kremlin that the West had found a way around that by fostering "colour revolutions", corroding and changing governments from within to pull them into its political orbit. The zero-sum mentality that characterized Soviet strategic thought during the Cold War told Russian leaders that the Western orientation of many formerly communist republics and the "colour revolutions" were the effects of Western hegemonic aggrandizement. Putin and his allies believed that as a great power with its own exceptional destiny, Russia was entitled to its "zone of privileged interest", an imperial space it controlled. For Russian nationalists like Putin, nation and empire were not discrete entities. The nation was the empire; the empire was the nation – the most obvious expression of that, the millions of "Russians" (including Russian-speakers who identified with other countries) left outside the *Russkiy mir* who, in Putin's mind, craved Moscow's protection.

These ideas seem to have developed gradually and there were deviations along the way because Putin was not the only person steering Russian policy. Dmitry Medvedev, for a while, pursued a subtly different path. The defence and foreign ministries didn't always see eye to eye. Sometimes, what outsiders took to be evidence of Russian strategic innovation was the Russian leadership trying to make sense of the world around them. Policy and practice are not developed in a vacuum: they are opportunistic, reactive. They draw on incomplete information, misperception and false assumptions. They are often ad hoc. In Georgia and Crimea, for example, although the Russian military clearly planned for invasion, the timing and circumstances were dictated by others. In Donbas, Russian strategy was chaotic and ad hoc. Thus, although we can tell a story about how Putin's worldview evolved towards Hobbesian war with the West, it is a necessarily imprecise and contingent story.

The story precedes Putin. As we have already seen, the idea that Russia should enjoy a privileged position within the former communist space was never entirely disowned when the Soviet Union dissolved, but Russia in the 1990s was in no position to act on its ambition. That changed in the 2000s. Putin's first two terms in office were dominated by Chechnya, but in government circles there was a growing sense of unease with the West caused by Western criticism of the war in Chechnya, humanitarian intervention in places like Kosovo, and the War on Terror – especially the 2003 invasion of Iraq. Putin saw these as part of a bigger pattern of hegemonic expansion that also lay behind the "colour revolutions" and the desire of countries to join NATO and the EU. As Putin saw it, the West never stopped fighting the Cold War but merely substituted the outward hostility of nuclear competition for the subterfuge of economic coercion, civil society activism, and democratization. Putin and his allies championed the *Russkiy mir* as an alternative to Western hegemony. That concept also predated his presidency, but Putin embraced it. In 2001, he spoke of a "Russian world" extending beyond the Russian Federation, beyond even Russian nationals living outside Russia to include Russian-speakers everywhere. That made Ukraine and Belarus not merely part of the Russian world but its core. It was Dmitry Medvedev who in 2009 established a presidential commission to prove the historical fact of shared (Russian) identity across the post-Soviet space.

These different strands of thought were brought together in Putin's 2007 Munich speech. Russia invaded Georgia the following year and that experience matured the idea of Hobbesian war. Georgia taught Putin that force could be used to good effect but that military modernization was needed. That began immediately and in 2010 Russian military doctrine identified NATO as the country's main rival. As he prepared to resume the presidency in 2012, Putin spoke of the need for Russia to embrace a wider range of foreign policy tools too. What he meant was the use of information and disinformation, cyber, subversion, economic coercion, assassination, proxies and mercenaries, to wage a war against NATO for power and influence but without direct military confrontation. The following year, Russia's Chief of the General Staff, Valery Gerasimov, argued that the line between war and peace had dissolved, and that Russia must be prepared to fight "hybrid wars" across the full spectrum of human endeavour to protect itself and its sphere of influence.

Gerasimov was responding to what he believed the West was already doing, not inventing new doctrine. "Hybrid war" is a term of art developed in the US, not Russia and the use of subterfuge to break the enemy's will is as old as war itself. The Soviet Union had dedicated whole institutions to spreading propaganda and disinformation in the West, Comintern chief amongst them. Russia's defence intelligence agency (GRU) was already in the business of foreign subterfuge, kidnapping and assassination. The key point is that the tactics were new but that by 2013, Putin and his allies believed themselves locked into the sort of permanent condition of war described by Hobbes. But whatever Putin believed, the West was not thinking in these terms. EU and NATO enlargement was driven by East European demand, not Western hegemonic expansion. Where Russia met resistance, it was the resistance of those defending themselves from Moscow not of Westerners bent on extending their hegemony. The more Russia pushed, the more the governments and peoples of Eastern Europe looked to the West for help. Thus, by acting as if it was locked in a struggle for power and influence with the West, Putin's Russia created the very thing it feared most: a competition for influence it could not win.

War by other means

As Putin explained in Munich, Russia's main strategic objective was to establish itself as a viable alternative to Western hegemony. Within that context, Russia's shadow wars have three principal objectives: (1) maintain control and political stability inside Russia's self-described sphere of influence; (2) limit external (i.e. Western) influence by supporting allies and discrediting rivals; and (3) project influence onto others, not just within Russia's sphere of influence but further afield as well. Shadow wars are ad hoc and opportunistic efforts to achieve these objectives by whatever means present themselves without resorting to armed force. They are conducted by state and non-state actors, and combine public and private motives. They are not always directed by the centre. Sometimes, the centre merely sets the course and allows others to innovate within regulated bounds. Deniability, plausible or otherwise, is key because the effectiveness of shadow wars often relies on the target's inability to detect or unwillingness to counter them.

We can distinguish non-violent from violent forms of shadow war. The non-violent try to take advantage of the connections between societies caused by globalization. As Mark Galeotti explains in *The Weaponization of Everything*, even though global connectivity has overall put downward pressure on the propensity of societies to wage war on one another, globalization did not eliminate competition between societies. In fact, it opened up new ways for societies to hurt one another. Interconnected societies are societies with any number of strings that can be pulled to achieve influence or control.

Economics. The most obvious non-violent way Russia has sought influence over others is through economic coercion. Russia has a major problem when it comes to using economics as an instrument of power.

The West's influence rests in part on the attractiveness of its economics. Other states *want* to join the EU, for instance, because they want economies and societies that look like EU economies and societies. This is "soft power" – a state or society's ability to persuade through attraction. Russia's problem is that its economic model is deeply unattractive. Russia's is a rentier economy that booms or busts according to the whims of natural resource pricing. There is little to gain beyond access to cheap gas and oil from tethering an economy to Russia's. Since Russia cannot attract others into its economic sphere, it must induce or coerce them. Putin regularly uses inducements to entice states into Russia's political orbit (especially the Eurasian Union), coercion to deter them from leaving, and sanctions as punishment for a wide range of transgressions.

Russia's principal form of inducement is energy pricing, a source of influence that helped bankrupt the Soviet Union. Russia creates dependency in Belarus, for example, by selling gas at well below market value, at a cost to itself of more than $70 billion since Putin came to power some economists estimate. From the Kremlin's point of view, the Nord Stream 1 and 2 gas pipeline projects are primarily about cutting Ukraine out of the gas supply chain and thus by depriving it of leverage creating energy dependency akin to that achieved with Belarus. Russia reduced the price Armenia paid for gas when it joined the Eurasian Union and bailed out Kyrgyzstan's debt-ridden pipeline network when it joined. The Kremlin encourages states to acquire huge debts to Gazprom that are then used as influence against them. This lever was pulled several times after the Orange Revolution in Ukraine for instance. In 2013, Russia used coercive threat and economic inducement to entice Yanukovych away from the EU association agreement. It tried the same with Moldova but with less success. Georgia was subjected to similar treatment before the 2008 war. Even friendly governments are not immune. In 2006–07, Gazprom increased the price of gas sold to Belarus causing a rift between Putin and Lukashenko, which served to remind the latter of his dependence on the former. Russia forced Belarus to sign over half-ownership of its pipeline network and acquired total control in 2011. Prices were threatened again in 2014 when Lukashenko proposed reinstating customs checks between the two countries and trading in dollars not roubles.

Beyond that, Russia has repeatedly used economic sanctions to induce, coerce and punish states that it seeks to influence. Both Georgia and Ukraine, for example, were hit with trade embargoes and visa restrictions before they were hit with military force. But economic coercion has failed as often as succeeded, and "success" has sometimes come at a huge cost. Belarus, Armenia and Kyrgyzstan may have been persuaded to join the Eurasian Union but it costs the Kremlin billions of dollars of lost revenue each year to keep them there. Their small economies and limited trade means that the customs union itself is of little economic value thus making the Eurasian Union a costly political project dressed up as an economic venture.

Russia was itself sanctioned by the West after its annexation of Crimea. Sanctions targeted a few dozen oligarchs closest to Putin and specific areas of the economy such as dual-use technologies, oil and gas exploration and exploitation, and banking. There is evidence they had some effect. Russia's GDP growth in the five years after 2014 averaged just 0.46 per cent, much lower than the 3.06 per cent of the five years before. But the oligarchs' pain was tempered by government subsidies, Putin spending more of the strategic reserve to protect his political base. Russia fired back with its own sanctions, banning the import of food and other goods from the West, causing shortages but also stimulating local producers. Russia also set about sanction-proofing its economy, greatly reducing its dependence on the US dollar and looking eastwards to new trading partners in Central Asia, Iran and China: the Eurasian dream. Western sanctions thus succeeded no better than Russian in creating political influence. The reasons why are complex and varied but seem hinged on the fact that although sanctions deepen divides between governments and peoples, leaders can respond with the rhetoric of "us versus them" blaming hardships on pernicious foreigners and rousing a "rally round the flag" effect. The ineffectiveness of these sanctions also suggests there are things people on both sides of the divide care about more than marginal changes to their lifestyle.

Cyber. The internet offers another medium for warfighting without military force. The Kremlin has employed cyberwarfare aggressively, the GRU, FSB and Foreign Intelligence Service (SVR) all conducting their own cyber operations. Cyberwarfare is also conducted by state-sanctioned troll farms, most famously the Saint Petersburg "Internet Research Institute" owned by Yevgeny Prigozhin, associate of Putin and financier of the private military company Wagner Group, of which more later. Troll farms conduct state- and self-directed cyber-attacks, the latter guided by broad parameters set by the state. There is also an army of self-directed "patriotic hackers" recruited and mobilized by the FSB, which sets their targets and lets them loose. It was these patriotic hackers that spearheaded a massive 2007 cyber-attack on Estonia. Likely triggered by Tallin's decision to relocate a Soviet war memorial statue, the cyber-attack targeted Estonia's parliament, government, financial sector, newspapers and television stations. Sometimes, war is waged in the shadows in support of more conventional warfighting. During the Georgia war, patriotic hackers attacked Georgia's critical systems and used Georgian sites to spread Russian propaganda. They did the same to Ukraine after 2014 and again in 2022. In 2018, Russian hackers targeted the opening of the Winter Olympics in Pyeongchang, likely retribution for the International Olympic Committee's decision to ban the Russian team because of its systematic violation of anti-doping rules.

Active Disruption. Propaganda and disinformation are major elements of Russia's shadow wars. RT beams the Kremlin's view of the world into foreign living rooms. Active disinformation campaigns spread untruths designed not so much to curry favour as to weaken outright opposition. For instance, Russian television and Russian diplomats lie about who shot down MH17. They lie about who gassed Syrian children. They lie about what happened in Crimea. Of course, every great power uses propaganda to create soft power for itself. The CIA and US Department of Defense used Hollywood to sell a political message that wasn't always completely true. But lying too often damages trustworthiness and risks a state's credibility and with it its capacity to influence, persuade, or attract others. Russian government figures have told so many lies to so many people that even their friends no longer find them believable. For example, the Chinese government didn't support Russia's position on MH17. Belarussian President Lukashenko rejected the "little green men" hypothesis and called the annexation of Crimea a "bad precedent" before being forcefully corrected. Figures like Lukashenko go along with the Kremlin because they are dependent on it, not because they believe what the Russians are saying. Commitment is therefore wafer-thin.

The active disruption of other people's governments is a step beyond. I have already catalogued a long train of coercive interference in the political affairs of Georgians, Ukrainians, Moldovans, Azerbaijanis, Armenians and Belarussians. But the Kremlin's shadow war against the

West has taken Russian disruption operations well beyond the former Soviet space, the objective being to disrupt and divide Western societies. This is part Machiavelli, to weaken Russia's main opponent by turning it in on itself, part ideological (the Comintern legacy), to discredit the Western model of politics, undermine its economic success, and make it less attractive to others.

Russian active disruption focuses on the provision of financial, political and strategic support to disruptive political parties in the West. The Kremlin's objective is not to support one or other political agenda, it doesn't distinguish far-right from far-left. In Italy, for instance, the Kremlin actively courted the leftist "Five Star Movement" and the rightist "Lega Nord", both sceptical of the EU. Putin makes a big public deal of his anti-fascism, yet his Kremlin channels support to the far-right Hungarian Jobbik party and German "Alternative for Germany" (AfD) party. The French far-right National Rally, led by Marine Le Pen, received a loan worth more than $12 million dollars from the small and little known First Czech Russian bank connected to the Kremlin. It is not Le Pen's praise that Putin values, but her disruptive anti-EU and anti-NATO policies. Like Putin, Le Pen wants a weaker EU and weaker NATO. Le Pen serves Russian interests even when she loses by sowing disruption, disquiet and factionalism in the heart of Western Europe. The Kremlin and its trolls supported the "Leave" campaign during the UK's referendum on EU membership for much the same reason.

The most notorious case was the Kremlin's support for Donald Trump's election in the US. It is not difficult to discern why Putin preferred Trump to Hillary Clinton. Trump was sycophantic towards the Russian leader, disdainful of America's European allies. Putin thought Trump an idiot and a Trump presidency likely to be chaotic if not ridiculous, weakening US alliances and undermining American soft power. A Clinton presidency by contrast would likely be competent and stable, Clinton taking a harder line on questions like Crimea and Syria than Obama had. The Kremlin oversaw the funnelling of funds from Russian oligarch Andrey Muraviev to a political action committee supporting Trump's candidacy. The GRU, meanwhile, hacked and leaked damaging Democratic Party emails. Prigozhin's IRA conducted a massive trolling operation, using 30,000 fake social media accounts to disseminate more than a million posts viewed more than 100 million times by American voters. The posts spread disinformation, including allegations of voter fraud and other crimes, about Clinton and the Democrats. Russia's patriotic trollers added to the volume of bile.

The effects of Russian political disruption are ambiguous. Far-right and left parties have unsettled Western politics, and support for populism has grown – notably in Hungary and Italy (where populism long predates Putin) – but the much-predicted tidal wave has not yet happened. Le Pen was soundly defeated by Macron, support for the German far-right has declined. Russian subterfuge no more got Trump elected than it got the UK out of the EU. The Americans and British managed that all by themselves and the effects were not wholly advantageous to Putin. Because Trump faced difficult questions about his ties to Putin, he was inhibited from taking policy positions favourable to Russia. Some sanctions were toughened, and the US proved a hindrance to Russian policy on Syria. Trump's hostility towards NATO and the EU damaged US ties with both but not irredeemably. And although Brexit meant the UK's break with the EU, in its search for a new foreign policy identity Britain became even more stridently hostile to the Kremlin. More generally, whilst the Western way of politics has lost some of its appeal, the Bush-era invasion of Iraq and the global financial crisis did much more to engineer that than Russian disruption operations.

Espionage and Assassination. Politics is a dangerous business in Russia and assassination a common tactic. As "Chechenization" recharacterized the war in Chechnya, assassination became a core Russian tactic – just as "targeted killings" featured large in the US War on Terror. One result was a wave of murders across Europe. Mamikhan Umarov, a Chechen critic of the Kadyrov regime, was shot in the head outside a Vienna shopping centre in 2020; Chechen blogger Imran Aliev was stabbed to death in Lille, his killer escaping back to Russia; Umar Israilov, one of Kadyrov's former bodyguards who testified about the systematic use of torture, was killed as he walked down a Vienna street; former Chechen fighter, Zelimkhan Khangoshvili, was shot dead in Berlin. In 2004, former Chechen president Zelimkhan Yanderbiev, was killed by a car bomb in Qatar. In 2016, two "Russian agents" were arrested in Turkey accused of shooting and stabbing another Chechen, Vahid Edelgiriev.

Sometimes the targets are in-house. Alexander Litvinenko, a former FSB agent, was poisoned in London by another agent, Andrei Lugovoy. Sergei Skripal, a former GRU operative, was poisoned in Salisbury by two GRU officers. Sometimes they are Putin's political opponents. Journalist Anna Politkovskaya and politician Boris Nemtsov are only two of the more than two dozen prominent political figures assassinated during Putin's term of office. In 2020, the FSB attempted to assassinate Alexei Navalny by poisoning him with the toxic nerve agent Novichok. Sometimes, the goal is to protect Russia's capacity to conduct clandestine operations. In 2014, elite FSB officers infiltrated across the border into Estonia and kidnapped an Estonian security officer who was closing in on an illegal smuggling ring used to channel funds into European bank accounts that could then be used to support covert operations without any evident financial connection to the Kremlin.

Montenegro. The different elements of Russia's shadow war came together in an audacious attempt to overthrow the government of Montenegro in 2016. The only part of Yugoslavia that chose to remain tied to Serbia when the federation collapsed in the 1990s, growing disquiet gradually turned to a push for independence. Montenegro finally seceded from Yugoslavia in 2006 and public opinion favoured membership of the EU and NATO. Montenegro's government set that course and thus became another political battleground. There was a strategic element too. Moscow wanted to secure access to a naval harbour in the Mediterranean to supplement its unreliable base in Tartus, Syria. Montenegro was the obvious choice, but the government in Podgorica rejected Russian advances.

From what we can tell, it appears the Foreign Intelligence Service (SRV) and GRU hatched a plan with Serbian nationalists and elements of Montenegro's pro-Serbian opposition to overturn the government. They seemingly planned to storm parliament on the night of national elections in October 2016 and declare victory for the pro-Serbian "Democratic Front". That would bring protestors onto the street and a group of plotters disguised as police would open fire on them whilst another assassinated Europeanist prime minister, Milo Djukanovic. The Democratic Front would claim the government was stealing the election and use the cover of a national state of emergency to claim power and reorient Montenegro away from the West. In the event, Montenegrin security services uncovered the plot a month before and two GRU officers, Eduard Shishmakov and Vladimir Popov, were named as being involved. Sources in Serbia suggest another 50 GRU officers were to have infiltrated Montenegro from there on election night to neutralize

Montenegrin special forces. The SRV was also implicated in the plot, through the Russian Institute for Strategic Studies, a supposedly "independent" institute staffed with former intelligence officers that had, among other things, run Russian interference in the US presidential election.

Assassinations and coup plots are not new features of world politics. Look beyond the technologies and characters and we see practices reminiscent of the Cold War. Back then, the Soviets often used criminals to infiltrate organizations and finance operations, for instance offering all kinds of support to radicals like the German Baader–Meinhof gang. Allegedly, one of Putin's functions in Dresden was to groom and manage individuals that could be used for information gathering and disruption in the West. The Soviets kidnapped and assassinated opponents and controlled how other governments behaved. Putin's Russia has nothing like the power the Soviet Union had but it does have the same institutions, people and training programmes – Putin is one of their products.

War by paid proxy: the Wagner Group

Russia's mercenary armies come in two forms. There are the real mercenaries, individual guns (and sometimes crewed tanks) for hire. When the Soviet Union collapsed and hundreds of thousands of soldiers were abandoned, some made a living doing what they do best. From Nagorno-Karabakh to South Ossetia, Transnistria to Tajikistan, the multiple wars of Soviet collapse were fought not just by new republics and restive secessionists but by moonlighting members of the Soviet Army. Veterans of one war popped up in other wars: Russians fought on both sides of the first Nagorno-Karabakh war; Chechens have fought on both sides of Russia's wars in Ukraine. These mercenaries were symptoms of collapse, sometimes useful to the Kremlin, sometimes not.

The second kind of mercenary, the kind that has come to dominate the scene since Russia's intervention in Syria, is altogether different. These are the mercenaries of private military companies – "PMCs". They are corporate employees, not individual guns for hire. Russian PMCs are only nominally private, for in practice they are instruments of the state. Much of their work is directed by the Kremlin and even that which isn't, the work they undertake to sustain themselves financially, is conducted within parameters set by the Kremlin. Indeed, the regulation of Russian PMCs is a perfect example of Putin's vertical of power in action. Formally, PMCs are prohibited by Russian law. Informally, they do the Kremlin's bidding. Operating outside the law gives the Kremlin deniability but more importantly keeps the sword of Damocles hanging over the mercenaries as a means of control. At any moment and for any reason the Kremlin might sweep them all into jail. This was the fate that awaited Vadim Gusev and Pavel Sidorov – leaders of a PMC known as the "Slavonic Corps" – in 2013 when they returned to Russia from Syria. The pair had recruited around 200 mercenaries to fulfil contracts for the Syrian government, apparently with the Kremlin's blessing since they travelled on government helicopters and used government bases. Once in Syria, they diversified into new contracts, including one guarding oilfields in the eastern governate of Deir ez-Zor. This contract may not have been with the Syrian government and some in the Russian government, most likely the FSB, were worried the mercenaries were overstepping the mark. They were withdrawn and Gusev and Sidorov sentenced to three years in jail, sending a clear message to the whole sector.

The most prominent but by no means only Russian PMC is the Wagner Group, so-called after the *nom de guerre* (all self-respecting prominent Russian mercenaries and volunteer soldiers have one) of Dmitry Utkin. "Wagner" because Utkin, a GRU Lieutenant Colonel who branched into the private sector on retirement, admires the Nazis. Wagner's roots are intentionally opaque, but it was a successor to the Slavonic Corps and a marriage between private sector and GRU officiated over by the group's financier, Putin ally Yevgeny Prigozhin. Utkin provides the muscle; Prigozhin the money, contracts and political direction. Like Putin, Prigozhin is a product of Saint Petersburg. He graduated from school into an unsuccessful life of crime, spending most of the 1980s in jail, convicted of robbery, fraud and other offences. Prigozhin was released into the death throes of the Soviet Union and, popular legend has it, set up a hot dog stall in Saint Petersburg. His empire grew from that one stall. Restaurants followed, including one of Saint Petersburg's finest. This sort of ascent could only have happened with the help of someone powerful. Someone like Vladimir Putin, the man managing business in the city at the time. When Putin became president, it was to Prigozhin's restaurant that he brought the American president. Prigozhin branched out into catering and cleaning, winning lucrative military contracts. Whenever the defence ministry expressed dissatisfaction with his work, the matter ended up in court and Prigozhin invariably won. He diversified into cyber trolling and then private military services with Utkin and the GRU, using a labyrinthine web of companies to broker contracts, raise funds and acquire assets.

Prigozhin seems to be the middleman, brokering contracts from the Kremlin and its allies, the latter within parameters set by Moscow. The operation receives cash and equipment from the Russian state. Wagner employees equipped with the latest Russian armoured vehicles played an active role in the battle for Debaltseve in Ukraine in January 2015. But the group also finances itself by taking other contracts or accepting payment in kind, such as lucrative gold and diamond mining operations. Whatever contracts it takes, however, Wagner's principal focus is on prosecuting Putin's shadow wars. Usually this involves provision of military support to authoritarian leaders, a type of operation that has taken Wagner beyond Syria into Sudan, Central African Republic, Mali, Mozambique, Burkina Faso and Madagascar. Sometimes, Wagner is employed against governments unfriendly to Moscow, as in Ukraine and Libya. Putin publicly acknowledged the group in 2018, the PMC shares a training facility in Molkino with the GRU's Spetsnaz, several Wagner employees killed in action in Syria received the Military Medal for Courage in Death, and Utkin received a medal for bravery from Putin himself. Nominally private, Wagner is actually an instrument of state power.

Wagner cut its teeth in Ukraine. Three hundred of its mercenaries were employed as "little green men" in Crimea and were then contracted to work in Donbas, where they provided VIP protection, reconnaissance and conducted special operations. As already mentioned, Wagner participated in the bloody battle of Debaltseve where Utkin himself was reportedly wounded, which may account for his bravery award. Independent Russian sources suggest Wagner employees fulfilled special Kremlin contracts to assassinate rebel commanders who refused to toe Moscow's line. In the so-called Luhansk People's Republic, Wagner employees forcibly disarmed the rebellious "Odessa" and "Russian Cossack" brigades.

Wagner operatives were withdrawn from Ukraine in the early spring of 2015 and by autumn were in Syria. The Kremlin wanted a low-risk war there, so instead of risking Russian soldiers sent Wagner. After fulfilling their initial contracts, the mercenaries branched out, accepting contracts from the Syrian government and oil companies to secure or retake oil facilities from IS. Wagner was at the vanguard of the Syrian government's first offensive to retake Palmyra in spring 2016. They were involved in the second offensive in early 2017 too, sustaining dozens of casualties. By now Wagner had between 3,000 and 5,000 mercenaries inside Syria, including large numbers of Kadyrovtsy Chechens and Ingush. Units designated as "IS Hunters" received bounties based on the number of extremists killed. It was apparently in pursuit of one such bounty that in February 2018 Wagner mercenaries accompanying Syrian government troops clashed with Kurdish Syrian Democratic Forces (SDF) allied with the US in Deir ez-Zor governate. Another account has it that the operation was engineered by the GRU to test American resolve since some in the Russian military were concerned about the growing size of the mercenary force and its inclination towards operating independently. Whatever the back story, the clash with the SDF developed into the shelling of a base housing US Marines. The Marines summoned air support with devastating effect. Accounts differ as to whether Wagner lost "dozens" or "hundreds" of employees. They agree that losses were significant, and that the Russian army kept out of it, advising the Americans only that the targets were not Russian soldiers – perhaps another exercise of informal regulation like the arrests of Gusev and Sidorov before.

Message conveyed; Wagner was soon back in action. Besides helping Damascus defend Deir ez-Zor, the mercenaries played active combat roles in the retaking of eastern Ghouta in late 2018 and in the 2019 and 2020 offensives on Idlib, where they encountered Turkish armed forces. As Russia wound down operations and redeployed forces from Syria to Ukraine, Wagner continued conducting limited operations against IS cells and supporting the Syrian government. By now Wagner was diversifying, using the capabilities built up in Syria to pursue a broader range of Russian foreign policy and private economic interests in Africa.

Each of Wagner's African operations is significant, but arguably the most significant is its role in Libya, a third front of Russia's clash with Turkey. Russia's Libya intervention is also a reminder that what matters most to the Kremlin is a foreign actor's orientation towards Moscow not its status as government or rebel or the type of regime it is. From 2017, the Kremlin gave political and military support to Field Marshall Khalifa Haftar's forces opposing the UN-recognized Tripoli government. Haftar is an old-school military strongman, who promised to defeat IS and the other Islamist extremists that took root in Libya after 2011. That chimed well with Moscow's aspirations, but Haftar – whose primary foreign sponsor is the United Arab Emirates (UAE) – also offered a way for Moscow to influence Libya's future direction and improve its status and strategic position, perhaps the latter in the form of a permanent military base on the Mediterranean – an objective pursued in Montenegro. It also cannot have been entirely coincidental that the Tripoli-based transitional government Haftar opposed was sponsored by Turkey and Qatar. Wagner's Libya mission began in earnest in 2017 and soon there were at least 1,500 (Turkish sources claim 2,500) Russian mercenaries operating alongside Haftar's "Libyan National Army". Prigozhin's cyber capabilities were marshalled in support.

When, in late 2019, Haftar began a concerted offensive against the Tripoli government, he did so with the support of Wagner infantry equipped with APCs and even a Wagner-provided airforce comprised initially of a dozen refurbished Su-22s and crews. As the fighting continued into 2020, this airforce morphed into an ambiguous Haftar–Wagner–Russian enterprise equipped with more modern MiG-29s and Su-24s delivered from Russian bases in Syria. At least two of the MiGs were shot down, both pilots were Wagner employees. Yet despite this support, Haftar's offensive ground to a halt on the outskirts of Tripoli in early 2020 and in late March, the Tripoli government led by Fayez al-Serraj, launched a fierce counter-offensive backed by Turkish drones, intelligence and naval missiles. As they had in Syria, and would do again in Nagorno-Karabakh, Turkish drones went head-to-head with Russian Pantsir anti-air missiles and fighter aircraft whilst on the ground Russian mercenaries fought Turkish proxies. According to some estimates, Turkey organized the deployment of more than 15,000 Syrian fighters into Libya's civil war. Once again, the Turkish-backed side fared better. Libyan government forces made steady gains until the summer and by the autumn an uneasy peace had been achieved, Haftar having fallen short.

Beyond Libya, Wagner's Africa operations focus on supporting autocrats favourably inclined towards Moscow and pursuing private economic interests compatible with the Kremlin's goals. Wagner typically seeks payment in the form of mining rights or other concessions. This economic function morphed from personal to public when Russia was hit with severe financial sanctions following its 2022 invasion of Ukraine. Its foreign currency holdings frozen by the West, the Kremlin looked to get around that by employing Wagner to supply the state with Sudanese gold. Wagner's role in Sudan began around the same time as its mission in Libya, although Russian PMCs had worked with the Sudanese government before. By the end of 2017, around 300 Wagner employees were performing functions such as training, supporting special operations directed at South Sudan and the troubled Kordofan regime, and protecting uranium, gold and diamond mines. The following year, Wagner secured new contracts to train and assist government forces in the troubled Darfur region. Wagner employees allegedly assisted the violent suppression of anti-regime protests in 2019 and when the Sudanese army ousted President Bashir, the Kremlin and Wagner deftly transferred their loyalty to the new military regime. In gratitude, the junta granted the Russian navy basing rights on the Red Sea and permitted the building of a naval logistics centre and dockyard. Prigozhin, meanwhile, acquired rights to a lucrative gold mine and in the midst of the Ukraine war, Wagner began illicitly transporting vast quantities of Sudanese gold to Moscow, paying neither tax nor royalties. On the ground, however, Wagner continued to work closely with Sudan's "Rapid Support Forces" (RSF), an elite group established by Bashir which spearheaded Sudan's brutal counter-insurgencies in Darfur and the Nuba mountains as well as operating in Libya and Yemen. Despite the change of government, the RSF continued to grow in size (partly thanks to its association with Wagner) and responded to the government's attempts to contain it by launching an armed bid for power in April 2023 that plunged Khartoum into violent conflict.

Not all the gold was flown out of Khartoum or shipped from Port Sudan. Some of it was driven across the border into the Central African Republic (CAR) and spirited out of Africa from there. Wagner has established itself in CAR as a key service provider to President Faustin-Archangw Touadera, whose authority extends little beyond his capital, Bangui. As in Sudan, Russian corporate and state interests acted in tandem. Russian diplomats to the UN persuaded the Security Council to grant exceptions to an UN arms embargo imposed when CAR's civil war had threatened to escalate into genocide. Wagner then exploited the new loopholes by sending in arms and "trainers" to support government forces. It established a semi-permanent training base for Touadera's special forces and Wagner personnel conducted joint patrols and other operations with the president's men. In return, Prigozhin's other companies acquired diamond and gold mining rights. By providing Touadera the military capacity he wants without the inconveniences of the human rights and civilian oversight insisted upon by the UN, Wagner's operations undercut UN peacebuilding efforts. They stand accused by UN experts of appalling abuses of human rights, including disappearances, torture and sexual violence. Thus, just as Russia's support for Haftar undercut Libya's UN-backed government in Tripoli, Russian action in CAR undercuts a UN strategy Moscow itself has endorsed in the UN Security Council.

In Madagascar, the Russians had a bet each way, giving political advice and Wagner's close protection first to the sitting president and then to at least two of the main contenders for the 2018 presidential election. It was a win-win situation. The island's outgoing president, Hery Rajaonarimampianina permitted a Prigozhin-linked firm to buy the national chromite producer, Kraoma. Chromium is a highly sought-after corrosion resistant metal. It was not long before Wagner employees appeared with the mines' security detail. By the time Rajaonarimampianina's electoral defeat became inevitable, Wagner was already providing advice and security to his rival, Andry Rajoelina.

In Mozambique, the following year, Wagner was called in to help the government defeat an Islamist al-Shabaab insurrection responsible for the forced displacement of tens of thousands of civilians. Two-hundred Wagner mercenaries equipped with attack helicopters were transported into the east African country by the Russian air force. After some initial success, the tide turned as the insurgents were reinforced by foreign IS fighters and inflicted a string of defeats on Wagner and Mozambiquan forces. Lacking situational awareness or experience combatting sub-Saharan African insurgencies, Wagner sustained significant losses (piecing together various media reports, it appears that at least one-fifth of its initial force was killed). Their Mozambiquan employers became dissatisfied with the company's poor service and terminated Wagner's contract. The Maputo government hired South African mercenaries to replace them.

Wagner's most recent fronts were opened in Mali and Burkina Faso. Since 2012, Mali's government has confronted an Islamist insurgency spearheaded by al-Qaeda affiliated Ansar Dine. The UN authorized a stabilization mission, MINUSMA, to support the government and help it resolve a long-standing dispute with its Tuareg minority. France also contributed. But the UN/French operations failed to achieve a decisive victory, whilst tensions between them and the Malian government over the latter's poor human rights record grew. International observers wondered why the UN was expending time and resources supporting an authoritarian, corrupt and abusive government. Leaders in Bamako questioned the value of such hesitant partners. In 2021, the government of Mali turned to Moscow and the Wagner Group, presenting Putin with an opportunity to displace France and the UN as Mali's principal backer, and Wagner an opportunity to have a second bite at Africa's Islamists. Since opportunities for enrichment in Mali were much more limited than elsewhere, this political dimension was more important. Up to 400 Wagner mercenaries equipped with helicopters and vehicles were despatched on Russian government aircraft, some taking up training roles with government forces and some deploying to an old French base in the ancient city Timbuktu from which they conducted combat operations. Prigozhin's cyber arm conducted an information campaign to discredit the UN and French and create the impression that Malians supported the switch to Russia.

Determined to avenge its loss in Mozambique and operating alongside an army notorious for abuses, Wagner waged a brutal counterinsurgency characterized by civilian massacre. In late March 2022, Wagner was involved in an attack on what they thought was an Islamist stronghold at Moura. After a brief firefight, the Russians and their Malian allies massacred approximately 300 civilians. Moura was not an isolated incident but part of a pattern of indiscriminate killing, forced disappearances, sexual violence and looting reported almost everywhere Wagner operates in Mali. The UN claims Wagner is responsible for creating a new displacement crisis there.

Wagner contributed to a coup in Burkina Faso in 2022. The background situation was like that in Mali: Burkina Faso's government, backed by an increasingly disinterested France critical of government abuses, was unsuccessfully combatting an Islamist insurgency. Lieutenant Colonel Paul-Henry Sandaogo Damiba urged his President Roch Kabore to change course and employ Wagner instead. Popular frustration at France and Kabore was stoked by a social media and public disinformation campaign orchestrated from Saint Petersburg by Prigozhin's troll farm. This helped build a groundswell of anti-government sentiment, which Damiba used to overthrow Kabore. People took to the streets of Ouagadougou in support of the new regime, some waving Russian flags. Within days, Wagner employees were visible in and around the capital.

What began as an ad hoc response to a particular need had evolved into an important instrument of Russian state power, one used to prosecute Putin's wars in Ukraine and Syria, challenge Turkish influence in Libya, and win political allies, military bases and access to natural resources across sub-Saharan Africa. All this was part of a larger quest for influence and status, an element of Putin's war against the West. Wagner isn't always successful. It was defeated outright in Mozambique, has taken a battering in Ukraine, and experienced significant reverses in Syria and Libya. It is too early to say how it will fare in Mali and Burkina Faso, but there is little immediate evidence to suggest it will be any more successful than the French and UN. By fighting in the shadows, the Kremlin limits its exposure to risks and costs, while taking advantage of opportunities to challenge the West and the UN as they arise. Material advantages include privileged access to Sudanese gold and Madagascan chromite. There are wider political advantages too. When the UN General Assembly voted on Russia's invasion of Ukraine in February 2022, 141 governments condemned it. Only four supported Russia: Lukashenko's Belarus could not do otherwise; Assad's Syria was another; the other two were the most Stalinist governments on the planet, Eritrea and North Korea. Thirty-four governments abstained, including all except one (Libya) of the African governments associated with Wagner: Burkina Faso, CAR, Madagascar, Mali, Mozambique and Sudan.

Through Wagner, the Kremlin is translating limited military engagement into political influence and material gain across a wide front. It is not just the West that Russia is challenging but the UN too. In Mali, CAR and Sudan, Russian strategy functions in opposition to collective UN interventions the Kremlin has itself endorsed. Whether this proves to be a strategic masterstroke or counterproductive folly remains to be seen, but the evidence thus far suggests the latter more likely. Wagner's patchy performance will likely blunt its military reputation over time and its litany of abuse will potentially provoke a backlash.

Putin's Russia has been waging war in the shadows for years, a war directed primarily at Putin's "main opponent", the US. It is a war waged at the margins, in the spaces where Russia can fight with limited risk and without provoking a military response that Russia can ill-afford. This is war as disposition, that extends beyond military force to economics, information and political murder. But it is a war nonetheless, and the West was bound to catch on sooner or later. Tragically, it was not until the quiet of dawn was pierced by the sound of missiles crashing into Kyiv on 24 February 2022 that the wool was finally removed from Western eyes. The second invasion of Ukraine was a long time in the making. It was crafted in the more than 20 years in which Vladimir Putin, his allies, servants and followers developed their bellicose disposition. Thomas Hobbes explains what happens when this disposition takes root, when political life is understood as war. In the famous Chapter 13 of his book *Leviathan*, first published in 1651, Hobbes wrote that in this condition, "nothing can be unjust" because there is no law to violate. "Force and fraud" become the "two cardinal virtues" such that "there be no propriety, no dominion, no mine and thine distinct; but only that to be every man's that he can get, and for so long as he can keep it". In short, what is right is whatever I believe to be right; what is mine is what I can take by force and fraud. Putin's march towards Kyiv exemplifies this disposition perfectly.

Further reading

Keir Giles, *Moscow Rules: What Drives Russia to Confront the West*. Washington, DC: Brookings Institution Press, 2019.

Ofer Fridman, *Russian Hybrid Warfare: Resurgence and Politicizarion*. Oxford: Oxford University Press, 2022.

Mark Galeotti, *The Weaponisation of Everything: A Field Guide to the New Way of War*. New Haven, CT: Yale University Press, 2022.

Sergei Medvedev, *The Return of the Russian Leviathan*. Cambridge: Polity, 2019.

Bettina Renz, *Russia's Military Revival*. Cambridge: Polity, 2018.

8
Ukraine II

The moment of Putin's failure was caught on a phone's camera. It was Friday 25 February 2022. The phone belonged to Volodymyr Zelensky, president of Ukraine. Rumours that he and his government had fled Kyiv were swirling, fanned by Russian propaganda and trolls, but standing in the lamp-lit street outside his presidential office, Zelensky issued a simple message of defiance. "The PM is here, the Party leader is here, the President is here. We are all here". He continued, "our military is here. Citizens in society are here. We're all here defending our country, our independence, and it will stay that way".

In that moment it was evident that Putin had failed to topple Ukraine's government and replace it with one more subservient to Moscow. In the months and years that followed, Russia also failed to win a decisive military victory capable of coercing Kyiv back into the Eurasian sphere. Russia's naked aggression and Ukraine's heroic defence finally spurred an appalled West into action. Ukrainian forces pushed the Russian army back from Kyiv, Mykolaiv and Kharkiv. The war continues but as I write, Ukraine has conducted a successful counter-offensive in Kharkiv, has retaken the one major city Russian forces did manage to take in its initial advance, Kherson, and has fought Russia's much vaunted spring offensive to a bloody standstill in Bakhmut. Russian losses are colossal. The US estimates that Russia has sustained more than 200,000 casualties. It has lost thousands of tanks, APCs and drones, dozens of helicopters and aircraft. Things have got so bad that in September 2022, Putin was forced to order a general mobilization of men of fighting age, pouring badly trained civilians into the grinder to replace the professional troops eliminated by Ukraine's defenders. Wagner recruited prisoners to send to the front and provided the vanguard for the assault on Bakhmut in spring 2023, its cavalier attitude to the welfare of its soldiers resulting in appalling losses. When Ukraine started to turn the tide there too, Prigozhin launched an excoriating public tirade against Putin and the Russian army, who he blamed for the failure – a telling sign that the Putinist alliance is fracturing under the weight of its own war. It is often difficult to know what is happening in the Kremlin, but as I write two things seem clear: Putin surely knows that he is losing the war in Ukraine and neither he nor anyone else amongst his top officials has the faintest idea what to do about it. Russia's failure thus far already means that it has lost Ukraine. One way or another, Ukraine's future will lie outside Russia's imperial orbit. As if to confirm that, on 23 June 2022 the EU granted Ukraine (and Moldova) formal candidate status. The question now is not what Ukraine's future orientation will be but how much time it will take to get there and how many Ukrainians will have to die and suffer.

This second invasion of Ukraine was long in the making. Ukraine has always had a special place in the imaginings of Russian nationalists but was also always just one part of a bigger picture, that of an imperial state leading the *Russkiy mir* and its Eurasian acolytes. This picture developed over time: Russia, the noble but wounded great power; the search for reclaimed strength through unity, the unity only a strong state and leader could provide; the realization and recognition of its status through military revanchism and control of a "zone of privileged interest"; the return to the pinnacle of world power exemplified by peer competition with the West. As I have already explained, these ideas predated Putin, but Putin and his myth-makers like Vladislav Surkov put them all together bit by bit, and Putin and his warmakers like Sergey Shoigu and Valery Gerasimov put them into practice. In 2022, the Kremlin's myths crashed into a wall of Ukrainian realities.

For most of the two decades these ideas and practices fermented, the West was asleep at the wheel. Annexation of Crimea and the downing of MH17 roused it, but it took all-out invasion to make the West realize what had been going on for years. That realization changed decades of policy. It was Ukrainians fighting and dying of course, but they did so increasingly with Western weapons and Western training, backed by swingeing Western sanctions and a determined effort to end dependency on Russian energy. Putin's Russia was now in a sense at war with the West because that was how Ukrainians saw themselves, and how more Europeans and North Americans than ever before now saw Ukrainians.

The steps to war

Russia's path to war in Ukraine needs to be understood in the context of Putin's past imperial wars. That story explains how Russia came to persuade itself that war was a legitimate and effective way of imposing control over its self-proclaimed sphere of privileged interests. The specific steps to the 2022 invasion are an interlaced story in which Putin, isolated by his own authoritarianism and fear of Covid-19, and a small cohort of loyal lieutenants convinced themselves that their struggle for Russia's Eurasian sphere was coming to a decisive point, that a divided and weak Ukraine was the fulcrum for that struggle, and that Russian military power and national will could outmatch those of a divided and chastened West. The first of those steps was Covid-19.

Vladimir Putin did not have a good pandemic. Officially, more than 376,000 Russians died with Covid-19 and more than 17 million were infected. That made it by far the most severely affected European country, and these figures likely under-report the true extent of the catastrophe because once it became clear in the early spring of 2021 that the authorities did not have the virus under control, they declared victory, stopped reporting Covid deaths, and changed the way they counted to include only those who died *because of* Covid. Per capita, Russia's infection rate was not unusually high, it was remarkably similar to Ukraine's, but it experienced an extraordinarily high mortality rate, double that of the US, more than double Britain's, and 20 times greater than Australia's. Only Peru's mortality rate was worse. These figures tell a story of a healthcare system in collapse, a system unable to test people and fatally incapable of treating them. Russian scientists did succeed in developing a reasonably efficacious vaccine, Sputnik V, but the state helped torpedo it. Russians are famous for their vaccine hesitancy. In its enthusiasm for its own vaccine, the Russian state systematically denigrated the foreign alternatives but the torrent of misinformation it unleashed undermined the confidence Russians had in their own vaccine. As a result, Russia's vaccine acceptance rates were the lowest in Europe. Far lower than in Western Europe and as much as 10 per cent lower than in Ukraine.

Putin himself seemed uncertain about the efficacy of Sputnik V. Whereas other world leaders jumped to the front of the vaccine queue to encourage their citizens, Putin hesitated. Uniquely among world leaders, Putin was not photographed having his vaccine, leading to speculation he was refusing to have it. In June 2021, the Kremlin announced the president had been privately vaccinated with Sputnik V but that didn't end the speculation. Some thought Putin had not been vaccinated at all. Others that he had received a Western vaccine. We can't be sure of the truth, but what we do know is that after an initial burst of activity where the president, sporting a Hazmat suit, toured a hospital he retreated deeper and deeper into self-isolation. By the second half of 2021, Putin had retreated so far that he rarely appeared in public or in a room with others. On the rare occasions when he did, Putin positioned interlocutors at the far end of ridiculously sized tables to enforce social distancing. Authoritarian leaders often become distanced from reality, Covid-19 widened the gap between Putin's world and everybody else's further still.

During his isolation-induced spare time, Putin wrote a 7,000-word essay "On the Historical Unity of Russians and Ukrainians". As a work of history, the essay has little to commend it but as a guide to the president's thinking it is invaluable. It is an astonishing essay that bears close reading. Its opening: Ukrainians are not a distinct people but are really Russians deceived by nationalists and the West. Its closing: Ukrainian sovereignty can only be realized through Russia. Statements of Russian imperialism pure and simple, direct (so we are to believe) from the president's hand. Putin set out his central thesis at the start: "Russia and Ukraine were one people – a single whole". "The wall that has emerged in recent years between Russia and Ukraine, between the parts of what is essentially the same historical and spiritual space" – the wall of separate statehood – "to my mind is our great common misfortune and tragedy". The tragedy of separation was "the result of deliberate efforts by those forces that have always sought to undermine our unity" – Westerners and their Ukrainian nationalist stooges. Modern Ukraine, Putin explained, "is entirely the product of the Soviet era" and the terms on which the Union dissolved were grossly unfair. Citing his former boss, Anatoly Sobchak, Putin argued that when states left the Soviet Union, they should have taken with them territory they entered the union with. Crimea and almost everything east of the Dnieper, in other words, should have been Russian. The final indignity was inflicted upon Ukraine's Russian population after independence. They had been "forced to deny their roots", to assimilate in an "ethnically pure Ukrainian state", acts "comparable in its consequences to the use of weapons of mass destruction against us".

Ukraine's political system is an "anti-Russian project" authored by the West. The "true sovereignty of Ukraine", Putin concluded "is possible only in partnership with Russia". According to a report by the *Vyortska* online newspapers, based on interviews with people close to Putin, the essay's first draft included an open threat to invade Ukraine unless Kyiv brought itself into line. The threat was reportedly deleted before publication, but if true shows where Putin's thought was heading.

Meanwhile, all was not well within the *Russkiy mir* and its surrounds. The year 2020 was one of elections in Belarus, the most reliable of Russia's partners, especially now Aleksander Lukashenko had been subdued by economic dependency. Most Belarussians, however, were tired of Lukashenko and would likely vote for change if they could. Minsk and Moscow worked on a "preventive counter-revolution". Sergei Tikhanovsky, a popular entrepreneur and YouTube activist, was arrested in May, accused of being a foreign agent bent on undermining the state. That brought protestors onto the street. More potential opposition candidates were arrested over the following months, including the most prominent leader, Viktar Babaryka. Lukoshenko claimed there was a Western plot to undermine Belarussian democracy. The authorities refused to allow the OSCE to send election monitors. But the pressure only grew. Sviatlana Tsikhanouskaya, wife of the jailed Tikhanovsky, announced her candidacy and other prospective candidates withdrew to give her a clean run at the dictator. Government crackdowns and daily protests continued. Security forces used tear gas and rubber bullets, protestors threw slippers.

Election day came in August amidst protests and violent suppression. Despite evident popular hostility towards the president, the authorities claimed Lukoshenko had won over 80 per cent of the votes. Signs of election fraud were everywhere, independent exit polling suggested Tsikhanouskaya won about 60 per cent of actual votes. Protests erupted across the country to which the security forces responded with their customary heavy-handedness. A dozen protestors were killed, more than 1,000 injured, and more than 30,000 arrested. Most were soon released but a few disappeared entirely amidst widespread reports of beating and torture. The following May, the Belarussian government hijacked a Ryanair jet flying between Athens and Vilnius, forcing it to land so they could arrest the journalist Roman Protasevich and his girlfriend Sofia Sapega. Putin announced the creation of a special "police" force that could be deployed into Belarus if needed to help restore order. The following month, he warned Belarussians that Russia would intervene if they pushed too hard. Russia supplied security advisers, weapons and other police equipment, and an economic bailout worth $1.5 billion to keep Lukoshenko in power. The Minsk government eventually throttled the protests, but it took immense effort over almost a year. For Putin, this was a reminder of Russia's struggle against Western hegemony and its insidious spread of "colour revolutions" and a shock since it revealed just how tenuous *Russkiy mir* and his Eurasian project were.

Another CSTO member, Kazakhstan, was also convulsed by crisis. In a context of authoritarian government, economic stagnation, endemic corruption and a poorly handed Covid-19 pandemic, a sharp increase in gas prices sparked a wave of protests in January 2022. More than 200 protestors and close to 20 security officers were killed and around 10,000 detained during the clashes. Uncertain about the state's capacity to maintain order, Kazakhstan's president Kassym-Jomart Tokayev invited the CSTO to intervene. Putin agreed, as did the CSTO's chairman, Armenia's prime minister Nikol Pashinyan – his own transformation from potential headache to compliant ally well on course. In 2014 Putin had spoken of Kazakhstan in similar terms to those he used for Ukraine, as a country beholden to Russia. Within days a force of 2,500 mainly Russian soldiers was deployed to protect key facilities such as Almaty airport and energy installations. With renewed confidence, Tokayev increased government repression and quelled the protests.

Things were also not going well for Putin in Ukraine. Putin's goal remained what it had been in 2014 but Minsk II had not given the Kremlin the leverage it wanted. The Russians insisted Ukraine legislate sweeping constitutional reform and federalize authority down to the regional level, awarding the LNR and DNR vetoes over Ukraine's foreign policy decisions, before progress on other aspects such as the removal of barricades, withdrawal of Russian forces, disarmament of separatist forces and restoration to Ukraine of control of its border, could be achieved. Poroshenko accepted that Minsk II required political reform but rejected federalization and insisted that the sequencing run in the opposite direction. The alternative, Poroshenko feared not unreasonably, would be a creeping annexation.

Unable to win leverage through Minsk II, Russia reverted to greater reliance on shadow wars. It continued to use Ukraine's energy dependency to extract concessions and drive a wedge between Ukraine and Western Europe. It also tried to make the annexation of Crimea a physical reality and began the creeping annexation of Donbas. Built at massive expense, the bridge across the Kerch Strait symbolized Crimea's physical integration into Russia. Following the path set in Abkhazia and South Ossetia, the Kremlin offered Russian passports to residents of Crimea and the two Donbas separatist entities. It also ratcheted up military pressure. There were regular flare-ups along the ceasefire line in Donbas and the Russian navy aggressively wrested control over the Sea of Azov from Ukraine, including in November 2018 by blockading, firing upon and capturing Ukrainian ships.

These efforts proved counterproductive. They undermined Minsk II and reinforced Kyiv's opposition to Moscow. A European arbitration ruling in 2017 found Gazprom's pricing manipulation breached competition law and awarded Ukraine's Naftogaz $2.5 billion. Work on the Nord Stream 2 pipeline continued but at a slower pace as the German government and business leaders debated the merits of economic cooperation with the Kremlin. Russian military pressure prompted countervailing measures. The EU, UN General Assembly and the International Tribunal of the Law of the Sea condemned Russian aggression. NATO expanded its joint manoeuvres with Ukraine in the Black Sea. The US and other NATO members began training Ukrainian armed forces and established a small office to coordinate training activities. Ukraine's association agreement with the EU came into effect on 1 September 2017.

There was a paradox at the heart of Russian strategy towards Ukraine. Putin's goal was to win over Ukrainians to his way of thinking. But with every move, he succeeded only in alienating them and uniting them in opposition to the Kremlin. Yet rather than change course, Putin persisted with coercion. Just how counter-productive this strategy was was revealed by the 2019 presidential and parliamentary elections. Putin tried reaching out directly to the Ukrainian people. He spoke repeatedly to the Ukrainian people, talking up their historic ties and promising an economic boon if they chose a pro-Russian government. To achieve that, he threw support behind Viktor Medvedchuk, a staunchly pro-Russian oligarch with close ties to Putin. The Kremlin threw its political technologists, media and money to support Medvedchuk and his Opposition Platform – For Life's presidential candidate, Yuriy

Boyko. To no avail. Boyko came in fourth, securing just 11 per cent of the vote. The resounding victor was an anti-establishment comedian turned politician, Volodymyr Zelensky, who soundly defeated his rivals, including President Poroshenko. It wasn't that Poroshenko had been a bad president, just that progress on reform had not been quick enough.

Zelensky's first act was to dissolve parliament and call new elections, presenting Putin and Medvedchuk a new opportunity to influence the government. A display of their political union dropped days before the election in the form of an Oliver Stone documentary posted to YouTube. "Revealing Ukraine" is a puff piece but its last few words are telling. In the final scene, Stone asks Putin: "Does he see these two countries coming together again?". Looking straight into the camera, Putin replies that "convergence is inevitable". Ukrainians disagreed. The Opposition Platform – For Life won just 13 per cent of the vote, some way behind the president's Servant of the People party which won 43 per cent. The Pew polling centre found only one in ten Ukrainians had confidence in Putin whereas eight in ten said they had no confidence in him.

Zelensky was less obviously pro-Western than Poroshenko. Part of his novelty was in not having a political platform. Politics should be a conversation, Zelensky believed. Putin thought Ukraine's new president weak and inept and assumed he could be bullied into concessions. When the two met in Paris, however, he discovered that Zelensky was as unbending as Poroshenko and so the Kremlin and its local allies ramped up the pressure. The separatists launched periodic spasms of violence that left 50 Ukrainian soldiers dead. Passportization proceeded apace, more than 600,000 Russian passports handed out in the two separatist entities. All this was calibrated to force Zelensky to relent on constitutional reform but instead they hardened the Ukrainian president's position and pushed him to seek closer ties with the EU and NATO. In February 2021, the Ukrainian government declared Medvedchuck a sponsor of terrorism for directing funds to the Donbas separatists and seized his assets. It was this, *Vyortska* reported, that prompted Putin to set a course for war. Since Medvedchuk was the vehicle Putin intended to use to control Ukraine by employing the techniques of shadow war, his political demise effectively closed that off and persuaded Putin a more direct approach was needed.

The three strands of the story came together in 2021: Putin, a president isolated by Covid-19 and developing imperialist fantasies;

a Russian state led by people who believed their country was a great imperial power locked in an existential struggle with the West; a shared belief in Ukraine's special place as the cradle of the *Russkiy mir* a growing sense that it was slipping away from them; and the failure of shadow war to nudge the situation in Moscow's direction. Russian strategy changed perceptibly towards the use of military coercion. A few weeks after Medvedchuk was arrested for colluding against the state, Russia deployed tens of thousands of troops close to the Russian and Belarussian borders with Ukraine for exercises. Additional Russian military personnel were deployed in Crimea and evidence pointed to military movements into Donbas too. By April, Russia had some 60,000 troops in and around Ukraine and a further 2,000 "advisors" in Donbas. Arms and ammunition were delivered, military overflights increased, and on at least one occasion Russian forces launched mortar fire at Ukrainian positions, killing four soldiers. Russia strengthened its Black Sea position by deploying vessels from other fleets and tried to prevent Ukraine's small navy from operating at all.

This was all counterproductive. Russian sabre-rattling forced Zelensky to become more strident, NATO responded with its own exercises, the US Navy deployed ships into the Black Sea, and the US and British air forces stepped up surveillance overflights. In late April 2021, Sergei Shoigu announced the completion of Russian exercises. Some Western analysts thought the April 2021 build-up a warning to the new Biden administration, but the principal driver seems to have been Russia's inability to coerce Zelensky by shadow war alone.

From April 2021 onwards we can see an inverse relationship between intent to coerce (by threat or limited force) and intent to control (by force). In April, intent to coerce remained higher than intent to control. Thereafter intent to coerce declined as threats failed whilst intent to control increased. Soon after Shoigu announced the end of the April exercises, the US reported that Russian military units remained in place around Ukraine's borders. In some cases, the personnel had withdrawn leaving their heavy equipment in situ. By October, the Russian military was on the move again, massing forces by Ukraine's border. The US confirmed in November that Russia had stationed about 70,000 troops there, the Ukrainian government put the figure closer to 100,000.

What made Putin believe outright invasion a prudent course of action? First, he thought Ukraine a weak and fragmented country unable to resist Russian power. The Ukrainian military had performed poorly in 2014 and the Russian military saw no reason to think anything had changed. Zelensky's popularity had quickly plummeted as his new form of politics failed to materialize. By December 2021, only a quarter of Ukrainians approved of their president. Putin also made the mistake of believing his own propaganda and assumed most Ukrainians thought as he did about the inevitable convergence of their two countries, discounting contrary evidence as the products of foreign manipulation. These beliefs were fed by the FSB, which ran a network of thousands of paid informants and collaborators in Ukraine reportedly costing more than a billion dollars. Headed by Sergey Beseda, the FSB operation confirmed Putin's judgement that Ukrainian society was on the precipice of collapse and that many if not most would welcome the Russians as brothers.

Second, Putin and his generals believed their own hype about Russia's military prowess. By 2020 Russia was spending more than six times more on its military than when Putin came to power. Much of that went into new weapons systems, creating an aura that the Russian military was again a peer competitor to the US. Decisive victories in Georgia and in Syria fed that belief – the latter especially so since it had been a Western-style intervention with few casualties. The Russian army had won in Crimea by the force of its presence alone, and Russian regulars had routinely outperformed the Ukrainian army in Donbas. Since Russia had not intervened directly in Nagorno-Karabakh, experience there could be safely ignored along with evidence from Georgia and Syria that some problems had not been resolved. Critical scrutiny would have cast doubt on this narrative of relentless success but authoritarian governments like Putin's tend to silence frank and fearless advice in favour of messaging that tells the leader whatever he wants to hear. Putin's Covid-related isolation turbocharged a tendency that was already there. No one seems to have demurred from the groupthink in Putin's circle about Ukrainian weakness and Russian military might.

Third, Putin assumed that the West would be incapable of meaningful resistance. On the basis that past behaviour is the best predictor of future behaviour, it is not hard to see how he might have reached that conclusion. Financial crisis, Brexit, and Trump had left the West riven with conflict and thus – Putin assumed – incapable of countering

Russia. Moreover, Putin believed Western society dangerously corroded by gender equality, sexual freedom and "woke" politics. For all Moscow's rhetoric about fearing NATO enlargement, Putin believed the alliance weak and fragmented and assumed it would not lift a finger to help Ukraine. More tangibly, the Kremlin judged that Russia's nuclear deterrence would inhibit the West from intervening directly. Whatever limited economic or military support it could provide would be too little and too late to prevent Russia achieving its objectives. Feckless Western efforts to arm Syria's opposition, the rapid collapse of Afghanistan's army in the face of Taliban, and the quick demise of the Iraqi army when IS invaded all pointed in that direction. Economic sanctions could be expected, but the pain could be tolerated and would be short-lived. Although Putin had spent a significant part of Russia's foreign currency reserves on the Sochi Olympics, pensions and military reform, the Russian government still held vast reserves thought to exceed $700 billion ($130 billion of which is held in gold) which could be used to cushion the blow of sanctions. States like Germany, Italy and Hungary were dependent on Russian energy and could be expected to temper Western sanctions.

Putin and his government thus assumed there was every reason to think that military force could be used to resolve their Ukraine problem and protect Russia's imperial space from external encroachment: Ukrainian resistance would collapse, the Russian military would overpower its enemy, and the West would stand aside. Every one of those assumptions was badly wrong.

Russia offered its terms on 17 December. That terms were offered to the US not Ukraine spoke volumes about Putin's worldview: Ukraine a mere pawn in a game between great powers. They required that NATO recognize Russia's imperium. Specifically, the parties "shall not create conditions or situations that pose or could be perceived as a threat to the national security of other Parties" (a tall order since Russia saw even rhetorical support for democracy as a threat), the US must not deploy military capability into NATO members that joined after 1997 without Russian consent, there must be no further enlargement of NATO, expressly no enlargement to include Ukraine, and NATO must not conduct military exercises in Ukraine, Eastern Europe, South Caucuses, or Central Asia. Russia was effectively demanding that NATO recognize an exclusive Russian sphere of military influence not just in the former

Soviet space but in Eastern Europe too. It was pure Yalta in an age where most of Russia's neighbours wanted Helsinki.

From what we can tell, it seems likely that the decision to invade involved three steps. First, the decision to switch from shadow war to military coercion, to develop an invasion plan, and forward deploy forces, was taken as early as February/March 2021. A preliminary decision to establish the military conditions to invade can be dated to April 2021, when forces and materiel were left in situ. The ultimatum can help us date the final decision to invade. Putin cannot have expected the US to accept these terms. To have even offered them suggests a decision to invade had already been taken since to walk back from the brink after the inevitable US-rejection would have been a great humiliation. This tells us that Putin had taken a final decision to invade before 17 December 2021. Most probably, it was taken even earlier, either before or during the October build up.

In January 2022, US intelligence reported Russia had up to 175,000 troops around Ukraine's border. This was no coercive show of force, the Pentagon insisted, but an invasion army. Wagner already had between 2,000 and 4,000 employees inside Ukraine, including approximately 400 infiltrated into Kyiv to assassinate 23 named Ukrainian leaders, including Zelensky, Prime Minister Denys Shmyhal, and the Mayor of Kyiv and former world boxing heavyweight champion, Vitali Klitschko. His ultimatum rejected, Putin had the Duma recognize the LNR and DNR as independent states in thrall to Moscow and addressing the nation in a televised speech on 21 February, blamed Kyiv for the impasse in Donbas, insisted Ukraine's NATO ambitions threatened Russian security, and reiterated that Ukraine was not a genuine state.

War

Putin announced his second invasion of Ukraine with an address to the nation broadcast in the early hours of 24 February. He explained that the US was intent on destroying Russia's traditional values and on imposing its own "false values". Western hostility was revealed, he said, by a litany of military interventions in Serbia, Iraq, Libya and Syria. Now, the West was bringing war to Russia's very borders, to its historic heartland of Ukraine. In Donbas, the West was using a perfidious Ukrainian government infested with Nazis to prosecute a genocidal war against Russians. "That genocide of the millions of people who live there". Russia would conduct a "special military operation" to "de-Nazify" Ukraine. Shortly after the prerecorded message was aired, missiles rained down on Kyiv and Russian forces crossed the border: from Belarus in the north heading for Kyiv seemingly intent on deposing the government in the expectation that Ukrainian resistance would collapse, from Russia in the northeast with Kharkiv as the main target, and from Crimea in the south heading west towards Kherson and Mykolaiv and apparently bound for Odessa if not Transnistria in Moldova (if maps revealed by Lukashenko can be believed), and east to take Mariupol and link up with separatist Donbas.

Things immediately went wrong. Russia's plan seems to have been to use paratroopers, special forces and Wagner mercenaries to enter Kyiv and kill or depose the government. The city that night was gripped with chaos and confusion, at least two kill-squads were repelled as they tried to get into the presidential compound. Urban legend has it that when US intelligence offered to spirit Zelensky out of Kyiv, the president replied "I need ammunition, not a ride". Whether or not he said exactly that, what mattered was that Zelensky stayed put. That act sealed Putin's failure. The president's staff barricaded windows and doors, his security detachments fought – and won – gun battles in the streets. All the while, Kyiv was rocked by missiles and the sound of gunfire to the north and west.

A pivotal battle occurred to the northwest of the city. A couple of hours after Putin's broadcast, Russian paratroopers carried by more than two dozen helicopters landed to seize the Hostomel cargo airport and its long runway 10 km from Kyiv city. The airport would have provided the Russians a way of rapidly transporting troops and weapons into Kyiv and was thus central to Putin's decapitation strategy. Although the airport was lightly defended, the Russians lost several helicopters. Nonetheless, the paratroopers that landed managed to briefly take control before a counterattack forced them to retreat and join up with a spearhead group of mechanized infantry moving ahead of a long convoy coming south from Belarus. Intense fighting continued, which, combined with the damage caused by combat and Ukrainian sabotage, meant the Russians could not use Hostomel as an aviation bridgehead.

The Russian mechanized troops that arrived in Hostomel were the advanced units of a 40 km-long convoy heading directly south towards Kyiv from Belarus, apparently intent on encircling Kyiv and strangling it into submission. Through a series of intense battles the column took Hostomel and Bucha to the north and northwest of Kyiv, but progress was painfully slow and losses significant. By the end of the first week of March, the convoy had almost entirely ground to a halt – more traffic jam than military offensive, its spearhead caught in intense urban fighting in Irpin and its rear subjected to Ukrainian assaults from the flanks. Russian forces used the sort of indiscriminate artillery fire employed against Grozny and Aleppo but lacked time and air superiority – Russian planes and helicopters were vulnerable to Ukrainian air defences. Russian equipment was poorly maintained, troop training and morale was low. Since poor maintenance meant they became bogged when they moved through fields, vehicles were forced to use paved roads restricting their advance. Communications were poor and easily hacked, captured Russian soldiers had outdated maps. Whole units sometimes got lost whilst others ran out of fuel and had to abandon their vehicles. All signs that Russia's military had been riddled with corruption: tyres not up to spec, machinery in a poor state of repair, troops poorly drilled. Hundreds of tanks, artillery pieces and APCs broke down and were abandoned or towed away by Ukraine's volunteer army of farmers. All the while the offensive-turned-traffic jam sat under the watchful gaze of Ukrainian drones and American satellites as Russian casualties soared.

This wasn't at all what Putin had expected. Far from fragmenting, Ukrainians across the country rallied to their government and exhibited a strong will to resist. To give just one example, Moscow had expected the city of Kharkiv, in Ukraine's northeast, to fall quickly. It is a short drive from there to the Russian border, less than half an hour and most of the nearly 1.5 million there have Russian as their first language. For Putin, Kharkiv was squarely part of the *Russkiy mir*, so he expected most people there to mobilize behind Russia. But that's not what the people of Kharkiv thought. Russian-speakers they might be, but they were Ukrainian not Russian, and they gave their allegiance to Kyiv not Moscow. The people of Kharkiv fought. National guards, ad hoc units pieced together by oligarchs and local organizers, and individual volunteers rallied to defend their city. They stopped the Russian advance on the city's outskirts and then the Ukrainian army pushed Russia out of the whole region the following autumn.

Far from shattering into a divided mess, the West finally woke up to the violent recidivist on its border. Its response was much more united, robust and (at the time of writing) sustained than anything Putin anticipated. Its first response was to impose further economic sanctions. The assets of Russia's central bank were frozen, meaning that it can't use the foreign currency it holds overseas (estimated to be up to $600 billion) to pay off debts; major Russian banks were cut off from the SWIFT international payments system; the EU and US imposed export restrictions denying Russia access to high-technology and other goods and expertise, including things like spare parts; oligarchs associated with Putin were subjected to targeted sanctions – Roman Abramovich was forced to divest himself of Chelsea Football Club; sanctioned individuals found their yachts impounded. As before, the Russian state tried to compensate the oligarchs for their losses, but its capacity to do so became more limited by the freezing of assets. Export bans will make it difficult for Russia to maintain efficiency in resource extraction, to maintain core technology and transport infrastructure, and repair and replenish losses in higher-end military hardware. Combined with a sharp fall in Russia's already low levels of foreign investment, sanctions over the longer term will likely end Russia's economic revival.

The Western response quickly moved beyond sanctions to military aid, especially once it became clear that Ukraine had both the will and wherewithal to resist effectively. NATO governments were careful to avoid direct intervention for fear of triggering Russian nuclear escalation, but instead furnished Ukraine with weapons, ammunition and training to a far greater extent than the Kremlin could have anticipated. At the time of writing, 39 states had provided military assistance to Ukraine, whilst more than 70 (including Azerbaijan, China and Serbia) had offered non-military aid. The US, UK, Poland and the Baltic states led the way on military assistance. For example, the US provided more than 200 tanks, 200 APCs, 100 self-propelled artillery pieces, 40 infantry fighting vehicles, more than 2,000 Javelin anti-tank missiles specifically designed for use on Russian tanks, 600 stinger missiles, and the Patriot anti-air system, as well as a wide range of ammunition, small arms, drones, jamming technology, and much else. Ukraine was also fielding US HIMARS systems – mobile guided missile launchers able to hit targets with precision up to 30 km behind the frontline – and more than 100 Leopard tanks provided by Germany and Denmark. The UK provided large numbers of Javelin and Starstreak missiles and more than 150 armoured vehicles as well as training to more than 4,000 Ukrainian soldiers. Poland supplied more than 230 tanks, 100 artillery pieces and much else besides. Turkey provided dozens of Bayraktar drones. In addition to the tanks, German military aid was hesitant but significant – representing a seismic shift in German thinking – more than 200 military vehicles, 500 stinger anti-aircraft missiles, and 30 armoured personnel carriers.

This was the tip of the iceberg beneath which lay a shift in Western Europe's security posture. Russia's invasion of Ukraine had finally convinced Europeans that the era of the post-Cold War peace dividend was over. Almost every major Western government committed to increase its defence budget, potentially plunging Russia into exactly the sort of financial spiral that had crippled the Soviet Union. German Chancellor Sholz committed to spending 2 per cent of GDP on defence, a level not achieved since 1991 (in 2020 Germany spent less than 1.5 per cent). This monumental shift will bring German defence spending to more than $52 billion annually, just $14 billion below Russia's. The UK and France also committed to increasing their defence budgets to around 2.5 per cent of GDP. With these shifts, British, French and German defence spending combined would more than double Russia's. But Russian actions have also reversed a long-standing tendency of Democratic Party presidents to want to cut American defence spending. The Biden administration added $29 billion to its defence budget in 2022, an increase that amounted to nearly half Russia's total budget – bringing US spending to over 12 times Russia's. Putin had taken Russia into a long-term military contest it could not possibly hope to win.

Before the invasion, Russia's border with NATO was a little more than 500 km long. Soon it will be close to 2,000 km long thanks to Finland's decision to join NATO. Sweden followed suit, a move that would effectively make the Baltic a NATO sea. Both had remained neutral during the Cold War. The invasion also brought European dependency on Russian gas into sharp relief. Although gas dependency caused some hesitancy on sanctions and some friction inside the EU, concerns about energy were outweighed by concerns about Russian aggression. Indeed, Russian aggression helped delegitimize arguments for restraint stemming from gas dependency. What is more, many European politicians found in the argument for weening themselves off Russian energy a compelling coincidence between geopolitical imperative and ecological necessity. Scholz committed Germany, Europe's largest importer of Russian energy, to stop buying Russian coal and oil by the end of 2022 and gas as soon as possible thereafter. The European Commission committed the EU to a similar path, away from Russian resources and towards a new mix of renewable energy. The long-term trend away from European purchase of Russian resources is already clear. Russian companies will be forced to find new markets by offering its goods at lower prices to less dependable buyers. Put that into a global perspective and all this adds up to Russia's long-term economic and military dependency on China.

On the frontline, Russia's initial "shock and awe" race to Kyiv soon gave way to something more familiar: war by atrocity. Indiscriminate bombardment to level cities that stood between the Russian army and its military objectives and the most horrific brutality to civilians caught up along the way. Men and boys tortured and summarily executed; women and girls tortured, raped and killed in their thousands. Horrors we first encountered in Chechnya now visited across Ukraine. The extent of Russian atrocities were exposed when a Ukrainian counter-attack in late March forced Russian troops out of Irpin, Bucha and Hostomel. The bodies of nearly 500 were recovered in Bucha alone. More than 400 had been shot at close range, their bodies still tied and blindfolded. Many showed signs of torture. More than 80 were female. The sex of five could not be determined due to their injuries. Survivors told horrific stories of how Russian troops went house to house looking for men and older boys. Of how they were beaten, tortured, bound and shot on the spot. Of how women and girls were taken as sex slaves, tortured and repeatedly raped, and then killed. Putin denied the Russians were responsible, Lavrov called the resounding evidence "fake".

Ukraine won the battle of Kyiv and forced Russia's northern frontline to retreat back to Belarus. Things initially went better for Russia in the south. Kherson, which controls entryways into Crimea, fell quickly. Russian forces bound for Odessa advanced on Mykolaiv but were stopped in the suburbs and pushed back towards Kherson. Meanwhile, Russian forces withdrawn from the north redeployed to the east. A strategic shift, the move was accompanied by a new rationale. Whereas in

February, Putin had insisted the objective was to "de-Nazify" Ukraine – that is, to change the government and bring Ukraine into Russia's political orbit – now he claimed the objective had always been to liberate Russian-speaking lands in Donbas and the south. Comparing himself and his mission to Peter the Great, Putin insisted "it is now also our responsibility to return [Russian] land". From May, the full force of the Russian army was brought to bear on Ukraine's east. Besieged Mariupol, however, still stood between Crimea and Donbas and the defenders held out to the bitter end. In March, Russian airstrikes there targeted the Donetsk Regional Drama Theatre, being used at the time as an air raid shelter. Six hundred civilians were killed when the theatre was brought down on top of them. The Russians claimed the theatre was a military target, then that the Ukrainians had blown it up themselves. In truth, this was just one of a long list of atrocities perpetrated by the Russian military. Mariupol was surrounded and pulverized with indiscriminate firepower in a war of atrocity taken straight from the Grozny and Aleppo playbooks. Conservative estimates suggest that 22,000 people were killed in the bombardment of Mariupol, nearly as many in two months there as in five years of siege in Aleppo. As more evidence comes to light of the scale of atrocities there, credible estimates of the dead have climbed as high as 50,000. Tens of thousands more were deported from Mariupol to Russia. The defenders held out at the massive Azovstal steelworks until late May. The last of them were members of the Azov Brigade, a far-right paramilitary group integrated into the Ukrainian army at the start of the war. Russia finally claimed what was left of Mariupol, an early war objective, only in late May and only after committing terrible atrocities and sustaining heavy casualties, losing possibly as many as 6,000 troops.

Russia's main effort was now focused squarely on Donbas. Overwhelming firepower helped it take territory gradually, but progress was painfully slow and as Western arms allowed Ukraine to fire back at range progress became increasingly expensive. Severodonetsk fell after a lengthy battle in June, neighbouring Lysychansk a couple of weeks later. Further north, however, the Ukrainian army pushed the Russians back from Kharkiv all the way to the Russian border in some places. By the end of summer, Russia's slow advance in the east had ground almost completely to a halt and in September, Ukraine conducted a stunning counter-attack that liberated Kupyansk and a swathe of territory in

Donbas. At the time of writing, Ukraine was continuing to advance in Donbas and had recently conducted another stunning offensive which succeeded in retaking Kherson. Russia meanwhile had ordered the chaotic mobilization of listed reservists, sparking a mass exodus and violent confrontations in Dagestan, and was buying military equipment from Iran and North Korea. Prigozhin, whose Wagner mercenaries had sustained thousands of casualties in Ukraine, was videoed trying to recruit in Russia's prisons. As winter set in, Russian strategy focused on the targeting of Ukraine's civilian infrastructure.

The British and American governments estimate that by August 2022 Russia had lost more than 1,000 tanks, 4,000 vehicles of all types, and 70,000 soldiers dead or wounded in Ukraine. All signs Russia was losing the war and growing desperate. Ukraine also suffered appalling losses, but it had no shortage of willing volunteers, was able to take advantage of training offered by NATO, and was becoming more adept at using the weapons the West was providing. With HIMARS, Ukraine has achieved the ability to target Russian arms dumps, supply points and command posts with precision. In military materiel at least, Ukraine's capacity was increasing whereas Russia's was being seriously degraded – and with little chance of refurbishment thanks in part to sanctions. To get a sense of Russian losses, consider the fact that they are already more than double the total losses of a decade of war in Afghanistan. Or that at sea the Russian Black Sea fleet has lost its flagship, *Moskva*, a warship *Saratov*, five patrol boats, a landing craft, and a tug to an enemy with no functioning navy. Or that in October, Ukraine disabled the Kerch Bridge.

At the operational and tactical levels, the Russian army was exhibiting all its old weaknesses. Intelligence failures meant things did not pan out at all as expected. Poor command and coordination meant the invasion's different elements rarely worked together. The Russian army's Battle Tactical Groups (BTGs) struggled to coordinate and manoeuvre together. Air and ground did not link up effectively – both problems that had been evident in Georgia. Ukrainians repeatedly succeeded in drawing Russian forces into urban battlefields neutralizing their superior firepower. Command and control was so poor that senior officers were forced to the frontline to direct things, presenting opportunities to the Ukrainians. At the time of writing, Russia had lost 14 Generals. The Russian military was not well stocked with its latest hi-tech equipment and had to rely on older material. Equipment failed and was poorly maintained – evidence that corruption remained rife.

Ukraine's war continues and it is impossible to say when or how it will end. What is clear, however, is that Putin will achieve none of his objectives and that Russia will emerge from the war in a far weaker condition than it started.

The end of Putinism?

The invasion of Ukraine was the culmination of two decades of Putinism. Both its causes and its failures emerged out of what had gone before. Whatever happens now, Putinism has been defeated in Ukraine. This is true irrespective of how and when the war ends or whether Putin clings on to power in Moscow. In all likelihood, he will cling on. Levada Centre polling shows most Russians support the war, although whether that support survives mobilization remains to be seen. There has been opposition. At the start of the war, the Russian government cracked down yet more on public dissent. Protesting or opposing the war became a criminal offence. More than 8,000 Russians were arrested for this in the first few weeks of the war. Media is more tightly controlled, opponents – especially in the resources industry – have died with even greater frequency in suspicious circumstances. Ravil Maganov, chairman of Lukoil, the one resource company to air public concerns about the war, plunged out of an upper-storey hospital window. Maganov was not the first senior Lukoil executive to die in strange circumstances in 2022. Over time, sanctions will compromise the livelihoods of ordinary Russians and may fracture the economic clauses of the social contract Putinism forged between state and people. An internal putsch is unlikely though. Like other long-serving authoritarians, Putin is surrounded by people who owe their standing and lives to him. Although figures thought responsible for Russia's military failures, such as Beseda, Shoigu and Gerasimov may be arrested, deposed or quietly shuffled aside, they are unlikely to move against the president for fear of bringing the house down upon themselves. Assuming it does survive, Putinism will become still less a consensual social contract and more old-fashioned authoritarianism-by-force.

But even if Putin survives, his vision of Russia as an imperial pole marshalling a sphere of interest in opposition to the American hegemon has been dealt a definitive blow. We can already see that. Only one member of the Eurasian Union, Belarus, supported Russia in the UN General Assembly. Armenia (dependent on Russian security guarantees), Kazakhstan (whose government had been propped up by Russian intervention only two months before) and Kyrgyzstan abstained (as did Azerbaijan, Uzbekistan and Turkmenistan). Georgia, Moldova and even Serbia voted with Ukraine in voting against Russia. As the war progressed Russia's imperial vice weakened further. Kazakhstan agreed to enforce Western sanctions against Russia, hoping to deepen cooperation with the EU whilst remaining within the Eurasian Union. Tokayev refused to join Putin in recognizing the LNR and DNR, instead affirming his support for territorial integrity and opposition to separatism. He also indicated that Kazakhstan would pursue a more independent path in future. Azerbaijan conducted fresh military attacks on Armenia and Nagorno-Karabakh, safe in the knowledge that Russia is in no position to fulfil its CSTO obligations. Further east, Tajikistan has done the same to Kyrgyzstan. Not even Belarus has proven altogether reliable. Sabotage of Russian military assets and transport infrastructure in Belarus has been widespread. Hundreds of Belarussians have volunteered to fight with the Ukrainians and there are reports the Belarussian military refused to assist the Russians. Despite crackdowns, the Belarussian opposition remains active. Putin's Eurasian project always suffered from Russia's soft power deficit – its inability to attract others to it. Russia's aggression in Ukraine has done it permanent harm.

War has finally caught up with the warmonger. Should Russia's imperial dreaming survive its battering in Ukraine, and it is by no means certain that it will, it will be as a Potemkin empire existing only in the minds of those who parrot its tropes.

Further reading

Andrey Kurkov, *Diary of an Invasion: The Russian Invasion of Ukraine*. London: Welbeck Publishing, 2022.

Luliia Mendel, *The Fight of our Lives: My Time with Zelenskyy, Ukraine's Battle for Democracy, and What it Means for the World*. New York: Simon & Schuster, 2022.

Samir Puri, *Russia's Road to War with Ukraine: Invasion Amidst the Ashes of Empire*. London: Biteback, 2022.

Samuel Ramani, *Putin's War on Ukraine: Russia's Campaign for Global Counter-Revolution*. London: Hurst, 2023.

Andrew Wilson, *The Ukrainians: Unexpected Nation*. Fourth edition. New Haven, CT: Yale University Press, 2015.

Chronology

19 October 1952	Birth of Vladimir Vladimirovich Putin
October 1956	Soviet intervention in Hungary
August 1968	Warsaw Pact intervention in Czechoslovakia
December 1979	Soviet invasion of Afghanistan
11 March 1985	Mikhail Gorbachev becomes General Secretary of the Communist Party of the Soviet Union
26 April 1986	Chernobyl nuclear disaster
February 1988	Nagorno-Karabakh autonomous oblast requests transfer to Armenia SSR
27 Feb–1 March 1988	Pogrom of Armenians in Sumgait (Azerbaijan)
February 1989	Soviet withdrawal from Afghanistan
April 1989	Nationalist protests in Tbilisi, Georgia
Aug–Dec 1989	Communist rule ends in East Germany, Poland, Hungary, Czechoslovakia, Romania, Bulgaria
November 1989	Putin confronts protestors in Dresden, East Germany
January 1990	"Black January" in Baku
May 1990	Putin appointed to Alexander Sobchak's administration in Saint Petersburg
August 1990	Armenia SSR declares sovereignty
December 1990	South Ossetia referendum on remaining within USSR
February 1991	Warsaw Pact dissolved
9 April 1991	Georgia declares independence from the USSR
19–22 August 1991	Attempted coup in the USSR
Aug–Sep 1991	Georgian national guard attempt and fail to take control of South Ossetia
1 December 1991	Ukraine referendum votes for independence from USSR
26 December 1991	Dissolution of USSR and establishment of the Russian Federation
Dec 1991–Jan 1992	Start of full-scale war between Armenia and Azerbaijan over Nagorno-Karabakh
24 June 1992	Ceasefire agreement for South Ossetia, deployment of Russian peacekeepers
Aug 1992–Sep 1993	War in Abkhazia; ceasefire agreement, deployment of Russian peacekeepers
Sep–Oct 1993	Moscow parliament crisis

Sep–Dec 1993	Georgian civil war
12 May 1994	Ceasefire agreement for Nagorno-Karabakh
December 1994	First Chechen War launched
June 1996	Yeltsin re-elected as President
August 1996	Chechens retake control of Grozny; peace settlement negotiated
Mar–Jun 1999	NATO intervention in Kosovo
7 August 1999	Start of the Second Chechen War
9 August 1999	Putin appointed Prime Minister of Russia
September 1999	Apartment bombings attributed to Chechen terrorists
31 December 1999	Putin appointed acting President of Russia
January 2000	Russia retakes Grozny
7 May 2000	Putin elected President
March 2003	New Constitution of Chechnya transfers authority to Kadyrov
November 2003	Rose Revolution in Georgia
14 March 2004	Putin re-elected to the presidency
Nov 2004–Jan 2005	Orange Revolution in Ukraine
December 2006	Major Russian combat operations in Chechnya largely over
8 March 2008	Putin becomes Prime Minister of Russia, Dmitry Medvedev becomes President
1–12 August 2008	Russian invasion of Georgia
Dec 2011–Jan 2012	"Winter of Discontent" in Moscow
February 2011	Arab Spring begins in Tunisia
4 March 2012	Putin elected to his third term as President
Nov 2013–Feb 2014	Euromaidan Revolution of Dignity in Ukraine
Feb–Mar 2014	Russian invasion and annexation of Crimea
Mar 2014–12 Feb 2015	Russian-backed separatist war in Donbas, Ukraine, and Russian invasion of Ukraine; concluded with Minsk II agreement
September 2015	Russian intervention in Syria
December 2016	Fall of Aleppo
18 March 2018	Putin elected to his fourth term as President
Feb–Mar 2020	Turkish intervention in Idlib, Syria. Russian operations wound down.
27 Sep–10 Nov 2020	Second Nagorno-Karabakh War
22 February 2022	Russian invasion of Ukraine

Index

9/11 46, 50, 67, 68, 110

Abashidze, Aslan 68, 69, 70
Abbott, Tony 113
Abkhazia 6
   and Russia 57, 58–9, 67,70, 71, 72, 75, 78, 100, 175
   and Russian invasion of Georgia 76–7,115
   conflict with Georgia 24, 60–62, 73, 75–6
   demography 23
   Russian peacekeeping in 65, 67, 74
Abramovich, Roman 33, 183
Adjara 23, 58, 67, 68, 69, 70, 71
Aeroflot 32
Afghanistan 6, 50, 179, 187
   Soviet war in 13, 14, 15, 28, 48
Aghdam 136
Akhalkalaki 71, 73
Aleppo 114, 117, 118, 121, 126, 182, 186
Aliev, Imran 157
Aliyev, Heydar 136–9, 141, 143, 145, 147
Alkhan-Yurt 45
"Alternatives for Germany" 156
Andropov, Yuri 14
Annan, Kofi 109
Arab Spring 88, 107, 108
Armenia 18, 71, 132–4, 135, 140–41, 143
   and EU 89, 92
   and Eurasian Union 64, 154, 189
   and Russia 130, 145, 153
   genocide (1915) 130
   in USSR 22–3
   see Nagorno-Karabakh

Asian Financial Crisis 33
Assad, Bashar al- 107–11, 114–22, 125–8, 167
atrocities 4–5, 9, 13, 22, 37, 43, 166
   in Bosnia 46
   in Chechnya 31, 44, 45, 47, 48, 49–50, 54, 72
   in Nagorno-Karabakh
   in Syria 116, 118, 121, 122, 124
   in Ukraine 182, 185, 186
Avturkhanov, Umar 31
Azerbaijan 6, 21, 22, 60, 66, 84, 129, 138–9, 140, 144, 159, 183, 189
   and EU 89
   and Eurasian Union 106
   and Russia/USSR 130, 132, 135–6, 137, 146
   And Turkey 141
   see Nagorno-Karabakh
Azov Brigade 100, 186

Babaryka, Viktar 173
Bagapsh, Sergei 71, 74
Baghdad 114
Baghdadi Abu Bakr al- 113
Bakhmut 169
Bahrain 107
Baku 22–3, 134, 135, 140, 147
Basayev, Shamil 32, 40, 47
   Budyonnnovsk siege 50
   in Georgia 60–61
   in Nagorno-Karabakh 134
   killed 54
Bashir, Omar al- 164
Batumi 71, 73

193

Bakiyev, Kurmanbek 69
Beijing 78
   Olympics 75
Belarus 64, 65, 69, 81, 89, 92, 134, 135
   and Russia 153–5, 167
   protests in 173–4
Berezovsky, Boris 33, 34, 39, 40, 41, 52
Berlin 3, 20, 78, 157
Beseda, Sergey 178, 188
Beslan, school siege 53–4
Bin Laden, Osama 46
Bitiyeva, Zura 49
Bosnia 46, 100
Boyko, Yuriy, 176
Black Sea 23, 52, 59, 60, 74, 75, 77, 81, 84, 94, 97, 115, 139, 175, 177, 187
Brezhnev doctrine 15, 18, 23
Brezhnev, Leonid 2, 13, 14
Budyonnovsk 32, 40, 50
Bucha 5, 182, 185
Bush, George W. 50, 67, 69, 74, 139, 157
Bucharest summit (2008) 74
Budapest 2, 19
Burkina Faso 161, 166, 167
Brazil 9, 114
Brexit, *see* United Kingdom
Buynaksk 41
Byzantine Empire/Byzantium 8, 81

Carter, Jimmy 14
Caspian Sea 115, 138, 141
Ceaușescu, Nicolae 12, 19, 20
Central African Republic 6, 161, 164
Channel One 33, 52,
Chechnya 4, 5, 6, 8, 21,
   constitution 55
   history 29–30
   nationalist movement 29–30
   first war 32, 37
   filtration 31, 49, 50, 131
   second war 39, 42–8
   insurgency 48–53, 66, 72
Chernenko, Konstantin 14
Chernobyl, nuclear disaster 16
Chernokozovo 49, 50
Chernomyrdin, Viktor 34, 40

China 9, 63, 78, 102, 106, 154, 183, 185
Chubais, Anatoly 57, 63
CIA 116, 155
Clark, General Wesley 46,
Clinton, Bill 45
Clinton, Hillary 156, 157
Cold War 3, 15, 150, 159
Collective Security Treaty 61, 64, 66
Collective Security Treaty Organization (CSTO) 64, 66, 129, 130, 139, 140, 142–5, 148, 174, 189
Commonwealth of Independent States (CIS) 61, 64, 67, 73, 83, 84, 85, 129, 135, 136, 137, 139
Congress of People's Deputies (USSR) 17, 23, 28
coup, attempted Soviet (1991) 12, 25–6
Covid-19 171–2, 174, 176, 178
Crimea 6, 9
   and Russia 172
   and Russian invasion of Ukraine 178, 181, 185, 186
   Gorbachev in 25
   history 82–3, 94–5
   Russian invasion/annexation of 8, 96–8, 100–102, 104, 105, 127, 149, 154, 155, 161, 170, 175, 177–9
Cedar Revolution (Lebanon) 69
crimes against humanity, *see* atrocities
Cuba 13
Czechoslovakia 15, 19, 20
Czech Republic (Czechia) 46

Dagestan 40–42, 50, 114, 187
Dagomys, peace agreement 59, 60
Damascus 110, 116, 121
Dar es-Salam 46
Daraa 114, 116, 122
Dayton, peace accord 46
Debaltseve 103, 161
Deir ez-Zor 116, 117, 160, 162
Denim Revolution 69
Djukanovic, Milo 158
Dnieper River 81, 82, 172
Donbas (Ukraine) 6, 82, 83, 84, 86, 94, 175, 176

INDEX

and Russian invasion of Ukraine
177–8, 180–81, 186, 187
war in 99–105, 150
Russian mercenaries in 161
Donetsk, 85, 86, 99, 100, 102, 103, 186
"Donetsk People's Republic" (DNR) 100, 102, 104, 105, 174, 180, 189
Douma 121
Dresden, Putin in 3–4, 37, 159
Dubrovka, theatre siege 43, 51, 52, 53, 54
Dudayev, Dzhokhar 30–32
Dyachenko, Tatyana 39

East Germany (German Democratic Republic) 3–4, 19–20, 46
Elchibey, Abulfaz 134–6
Egypt 88, 107, 110, 115, 120
Elistanzhi 45
Estonia 18, 26, 85
and NATO 105
Russian attacks on 101, 155, 158
Eurasian Economic Union 64, 89
Eurasian Economic Community 64, 85
Eurasian Union 64, 85, 89, 92, 106, 129, 139, 140, 142, 153, 155, 189
European Court of Human Rights 49
European Union 64, 130, 150, 151, 153, 156–8, 160, 166
and Armenia 140
and Azerbaijan 139, 143
and Georgia 70, 75, 78
and Ukraine 84, 85, 86, 87, 91–5, 103–107, 170, 175–6, 183, 185
Neighbourhood policy 89, 92
Erdoğan, Recep Tayyip 118–21, 123–7, 130–31, 141, 147–8

Fedayeen 131, 133
"Five Star Movement" 156
France 63, 74, 93, 95, 107, 137, 166, 167, 184
FSB 34, 39, 49, 53–5
apartment bombing allegations 41–2

Gadhafi, Muammar 88, 107, 110
Gagra (Georgia) 61

Galeotti, Mark 152, 168
Gamsakhurdia, Zviad 24, 58, 59, 60, 61, 62
Ganje, military base 134, 136
Gazprom 33, 52, 91, 153, 175
"Gerasimov doctrine" 149
Gerasimov, Valery 151, 170, 188
Georgia 6, 9, 18, 24, 25, 69, 84, 189
and EU 78, 89
and NATO 73, 74, 75, 91, 106
and Russia 66, 67, 71, 72, 138–9, 158
and Collective Security Treaty/Organization 66
conflicts in 59–60, 61–2, 75
constitutional structure 21
National Guard 24, 58, 60, 61
peacekeepers (Georgian) 70
peacekeepers (Russian) 70, 130,
Russian invasion of 8, 57–8, 76–8, 150–51, 155
see Abkhazia, Adjara, Rose Revolution South Ossetia
Germany 29, 74, 88, 93, 95, 101, 103, 105, 156, 179, 184
Ghouta 110, 119, 121, 162
Girkin, Igor ("Strelkov") 100
Glazyev, Sergei 93, 94, 99
Gongadze, Georgiv 86
Gorbachev, Mikhail 4, 12–15, 24–7, 30, 38
Glasnost 15–17
foreign policy 20
Perestroika 14–17
Gori (Georgia) 76, 77
Grape Revolution (Moldova) 69
Grazyev, Rahim 135–6
Greene, Samuel 7
GRU (military intelligence) 49, 54, 151, 154, 156, 157, 158, 160, 161, 162
Grozny 30–32, 43, 44–5, 47–9, 55, 182, 186
GUAM 66, 84
Gusev, Vadim 160, 162
Gusinsky, Vladimir 33, 34, 47, 51, 52
Gyumri, military base 130, 140, 145
G7/G8 101, 106
G20 111, 113

195

Haftar, Khalifa 163, 165
Hama 114, 116, 119
Hashimov, Polad 143
Hezbollah 107, 126, 127
Hitler, Adolf 18
　leaders likened to by Russian media 79, 83
Hollande, François 103
Holodomor 83
Homs 114, 116, 119
Honecker, Erich 19, 20
Horvath, Robert 87
Hostomel 181, 182, 185
Human Rights Council *see* United Nations
Hungary 12, 15, 19, 20, 157, 179
　and NATO 46, 105
Huseynov, Surat 135-7

Idlib 114, 115, 116, 119, 120, 121, 122, 123-8, 147, 162
Irpin 182, 185
India 9, 106
Indonesia 9
International Labour Organization (ILO) 28
International Tribunal of the Law of the Sea 175
Iran 107, 114, 119, 122, 127, 144, 147, 154, 187
Iraq 50, 69, 75, 77, 85, 108, 109, 150, 157, 180
　IS in 112-13, 179
Islamic State (IS) 107, 112, 115, 118
Israel 108, 120, 122, 140
Ivanov, Igor 68
Ivanov, Sergei 72, 96

Jabrayil 144
Jackson, General Sir Michael 46
Jobbik 156
Jinping, Xi 114
Jordan 107, 108, 122
Judah, Ben 8

Kabul 2
Kadyrov, Ahmad 54

Kadyrov, Ramzan 37, 54, 55, 72, 157,
Kadyrovtsy, 54, 55, 56, 162
Kaspiysk (Dagestan) 50,
Kazakhstan 30, 64, 174, 189
　and Eurasian Union 89, 92
Kelbajar 136, 137
Kerch Strait bridge 105, 175, 187
Kerry, John 111, 117, 118
Key West 138, 139
KGB 2-4, 14, 25, 34, 37-9
　see FSB
Khan Shaykhun 124
Khangoshvili, Zelimkhan 157
Khasavyurt, peace accord 39
Khattab, Ibn al- 40
Kharkiv 95, 99, 100, 169, 181, 182, 186
Khelmnytsky, Bohdan 82
Kherson 169, 181, 185, 187
Khmeimim, air base 115
Khodorkovsky, Mikhail 33
Khojaly, massacre 133, 134
Khrushchev, Nikita 15, 30, 83
Kirill, Patriarch 109
Kitovani, Tengez 60
Kocharyan, Robert 132
Kohl, Helmut 4
Kokoity, Eduard 67, 70, 74
Kosovo 35, 46, 150
　independence recognized 73, 74, 88
　see NATO
Kovtun, Dmitry 39, 72
Kravchuk, Leonid 26, 83, 84, 92
Kriuchkov, Vladimir, 25
Kurds 118, 120, 123
Kyiv 8, 17, 81-2, 86, 91, 95, 97-9, 100
　Battle of 5, 6, 149, 168, 169, 180-82, 185
Kyivan Rus 81, 82
Kyrgyzstan 64, 69, 88, 153, 154, 189
Kuchma, Leonid 84, 85, 86, 92, 94, 99

Lachin 134, 144, 145, 146
Latakia 114
Latvia 18, 21, 26
　and NATO 85, 105
Lavrov, Sergei 74, 77, 96, 111, 117-18, 120, 139, 141, 185

Le Pen, Marine 156
Lebanon 69, 108
Lebed, Alexander 32, 39
Lega Nord 156
Lenin 1, 15
Levada Centre 7, 188
Libya 107
   and Russia 130, 142, 165, 167, 180
   civil war 110
   NATO intervention in 88, 109, 111, 112
   Russian mercenaries in 6, 161, 162–4
Lithuania 12, 18, 25, 26, 82, 85, 105
Litvinenko, Alexander 39, 42, 79, 157
Luhansk 86, 99, 100–103, 161
"Luhansk People's Republic" (LNR) 100, 105, 174, 180, 189
Lugovoy, Andrey 39, 72, 157
Lukashenko, Alexander 65, 69, 88, 91, 153, 155, 157, 173, 181
Luzhkov, Yury 41

Madagascar 161, 165, 167
"Madrid principles" 138
Malaysia, Malaysian Airlines 101
Mali 6, 116, 141, 161, 166, 167
Maraga, massacre 134
Mariupol 99, 100, 181, 186
Martakert 136
Mashkadov, Aslan 39, 40, 54
Mazowiecki, Tadeusz 19
Mediterranean Sea 198, 117, 158, 163
Medvedchuk, Viktor 175, 176, 177
Medvedev, Dmitry 72–5, 76, 78, 88, 89, 90, 101, 107, 139
Melon Revolution (Kyrgyzstan) 69
Memorial 49, 50
Merkel, Angela 74, 101, 103
MH17 101, 113
Minsk 173, 174
   Peace agreements I and II 103, 104, 175
"Minsk Group" 137
Moldova 6, 9, 65, 66, 69, 84, 88, 89, 100, 106, 115, 130, 155, 170, 181, 189
Moltenskoi, General Vladimir 50
Montenegro 158, 163

Moscow 33, 34, 38–9, 63, 74, 76 ,77, 126, 164, 173
   August 1991 coup 25–6
   elite 28
   local government 17, 41
   murders 72
   public opinion 44
   terrorist attacks 50–51, 53
Mosul 112
Mozambique 161, 165, 166, 167
Mubarak, Hosni 107
Muscovy, 82
   Grand Dukes of 8
Mutalibov, Ayaz 131–4, 136, 137
Mykolaiv 169, 181, 185

Naftogaz 175
Nagorno-Karabakh
   and Armenia 132, 136, 143
   and Russia 135, 146,
   first war 130, 131, 133–7
   in USSR 21–2
   Operation Koltso 23
   Russian peacekeepers in 6, 130, 137–8
   second war 143–6, 147, 148
Nairobi 46
Najibullah, Mohammed 13
Nakhchivan 136, 142, 146
NATO 44, 64, 87, 156, 157
   and Azerbaijan 139
   and Georgia 69–72, 73, 74
   and Ukraine 84, 90–91, 175, 176, 177, 183
   as Russia's rival 150–51
   enlargement 5, 46, 68, 88, 105, 158, 184, 187
   intervention in Kosovo 35, 46
   intervention in Libya 107, 109, 110
   nuclear weapons 149
   Partnership for Peace 84
Navalny, Alexei 89, 158
Nazis 1, 5, 7, 64, 83, 98, 160, 181
Nemtsov, Boris 34, 40, 49, 158
Nizhny Novgorod 34
NKVD 1
Nord Stream 88, 91, 153, 175
North Ossetia 23, 24, 53, 59

Novye Aldy 49
NTV 33, 47, 51, 52
Nuli 75

Obama, Barack 88, 101, 110, 114, 156
Odessa 100, 161, 181, 185
Orange Revolution (Ukraine) 69, 70, 71, 87, 88, 90, 91, 94, 96, 153
  background 83, 85–6
Organization for Security Cooperation in Europe (OSCE) 173
  mediation in Nagorno-Karabakh 131, 137, 148
  observers in Chechnya 32
  observers in Georgia 75
  Ukraine ceasefire 103
Organization for the Prohibition of Chemical Weapons (OPCW) 122

Palmyra 114, 162
Paris 39, 113, 176
Pashinyan, Nikol 142–5, 147, 174
Patrushev, Nikolai 41, 96
Pavlov, Valentin 25
Plokhy, Serhii 26, 35, 106
Poland 12, 19, 20, 46, 95
  assistance to Ukraine 183
  and NATO 46, 105
Polish-Lithuanian Commonwealth 82
Politkovskaya, Anna 49, 56, 72, 158
Popov, Vladimir 158
Poroshenko, Petro 100–102, 174, 176
Poti 59, 60, 62, 77
Prague 2, 19
Prigozhin, Yevgeny 154, 156, 160–61, 163–7, 170, 187
Primakov, Yevgeny 34, 25, 41, 46, 47
Prokhorov, Mikhail 89
Protasevich, Roman 173
Putin, Vladimir
  attitude towards Ukraine 5, 81, 172–3
  Chechnya 43–7
  collapse of USSR 3–4, 11, 37–8
  Covid-19 170–71
  early life 1–2
  economy 66
  elections 47, 52, 89
  geopolitical vision (Yalta) 69–72, 87, 105, 108–109, 149–52, 168
  imperialism 6, 8, 24, 65, 85, 92, 104, 148
  KGB/FSB 3–4, 37–9
  Kursk 52
  Munich Security Conference speech 72–3, 105, 151–2
  Nagorno-Karabakh 147–8
  nuclear threats 98
  objectives in Ukraine 97, 104
  oligarchs 52
  "power vertical" 53–4
  presidency 4
  return to presidency 88–90
  prime minister 40
  second term 73, 88–9
  Saint Petersburg 38–9
  Syria intervention 114–15, 117
  Syria "peacemaking" 120
  relationship with Armenia 145
  relationship with Turkey 118, 125–7
  Ukraine, rationales for war 177–9
  warmonger 9
Putinism 7–8, 11–12, 55–6, 78–9, 87–90, 104, 188–9

Qatar 107, 108, 110, 157, 163

Raqqa 113
Reagan, Ronald 14, 15
Red Army, *see* Soviet Union armed forces
Red Army Faction 3
"Revolution of Dignity" (Euromaidan, Ukraine) 69, 87–95, 99, 100, 107, 112
Robertson, Graeme 7
Roki tunnel 75, 76
Romania 12, 19, 20, 105
"Rose Revolution" (Georgia) 57, 58, 68, 69, 85, 87
Rosneft 96
Rousseff, Dilma 114
Russia Soviet Socialist Republic 26
Russian Federation (Russia) 37, 78
  armed forces 31–2, 47–8
  Duma 70, 97

economy 12, 27–8, 34
Federation Council 97
governance 53–4
imperial history/revival 8, 21, 63
independence 27
Ministry of Internal Affairs (MVD) 54
political crisis 28–9, 46
Russkiy Mir 24, 62, 65, 67, 79, 92, 106, 150, 151, 170, 173, 174, 177, 182
Rutskoy, Alexander 28–9
Ryazev, Dadash 136

Saakashvili, Mikheil 58, 68, 69, 70, 79
and Russian invasion of Georgia 75–7
Saint Petersburg (Leningrad) 1, 2, 3, 8, 17, 37, 38, 44, 111, 154, 160
Saint Volodomyr/Vladimir 8
Samara 51
Samashki 31
Sapega, Sofia 173
Saraqib 125
Sargsyan, Serzh 139, 140, 142
Sarkozy, Nicolas 77
Saudi Arabia 107, 108, 120
Scholz, Olaf 185
Sechin, Igor 96
Second World War ("Great Patriotic War") 1, 8, 46, 72
Security Council, *see* United Nations
Serbia 46, 73, 158, 180, 183, 189
Sevastopol, 84, 91
Shamonov, General Vladimir 48
Shevardnadze, Eduard
foreign minister USSR 15
President of Georgia 58–61, 66, 67, 68
Shishmakov, Eduard 159
Shoigu, Sergey 98, 170, 177, 188
Shusha 133, 134, 145, 146
Shushkevich, Stanislav 26
Sibneft 33
Sidorov, Pavel 160, 162
Simferepol 97
Sisi, Abdel Fattah el- 110, 120
Skripal, Sergei 157
Skuratov, Yuri 39
Sloviansk 99, 100

Sobchak, Anatoly 37, 38, 39, 172
Sochi 59, 119–20
Winter Olympics 66, 73, 90, 95, 179
South Africa 9, 114, 165
South Ossetia 6, 67
and Russia 71, 72, 73, 100, 115, 159, 175
and Russian invasion of Georgia 76–8
conflict with Georgia 23–4, 57, 58, 59, 67, 70–71, 75
demography and history 60
Russian peacekeepers in 60, 62, 65, 74
South Sudan 6, 164
Soviet Union (USSR) 2, 6, 13–18, 108
armed forces 14, 20, 22–3, 24–6, 38, 147
Black Sea Fleet 84
collapse of 4, 10–11, 21, 26–7, 30, 58, 61, 63, 83
constitutional structure 17, 21
Communist Party of 13, 17–18, 28
economy 14, 16
espionage 159
Nagorno-Karabakh 131, 147
successors to 64–5
Union Treaty 26
Srebrenica 46
Stalin 1, 2, 15, 17, 18, 21, 30
Stavropol Krai 51
Stepanakert 133, 134, 144, 146, 147
Stepashin, Sergei 40, 42
Sudan 161, 164, 167
Sukhumi 60, 61, 74
Sumgait, pogrom 22
Surkov, Vladislav 85, 87, 89, 170
Syria 6, 141, 142, 144
and Russia 9, 69, 108–109, 112, 157, 158, 163, 167, 178
and Arab Spring 88, 110
and private military corporations 160–62
Chemical weapons use 155
civil war 111, 113
peace process 120–22
Russian intervention in 114–17, 118–19, 123, 124–8
Syrian Democratic Forces (SDF) 162

Taftanaz 125
Tajikistan 64, 106, 159, 189
Tartus, naval base 108, 115, 158
Tatars 30, 82, 83
Tatarstan 30
Tbilisi 23, 24, 59, 68, 77
Terek, River 45
Ter-Petrosyan, Levon 132, 133, 135, 142
Tikhanovsky, Sergei 173
Tikrit 112
Tilly, Charles 1
Tokayev, Kassym-Jomart 174, 189
Transnistria 6
Trump, Donald 9, 119, 120, 156, 157, 178
Tskhinvali 58, 59, 70, 74, 75, 76
Tsikhanouskaya, Sviatlana 173
Tulip Revolution (Kyrgyzstan) 69
Turkey 32, 70, 71, 74, 131
    and Armenia 132, 140–41
    and Azerbaijan 130, 143–4, 147
    and Libya 163
    and Russia 108, 127–8, 142, 162
    and Syria 107, 118–21, 123–6
    assistance to Ukraine 184
Tymoshenko, Yulia 90–93

Ukraine
    and CIS 83, 84
    and EU 84, 92–3, 170
    and NATO 74, 91
    corruption 92
    defence of Kyiv (2022) 181–2
    economics 84, 91, 102–103
    gas supplies 153, 174–5
    history 81–3
    Holodomor 83
    independence referendum (1991) 26
    military assistance to 183–4
    nuclear weapons 83–4
    Russian atrocities in 185–6
    Russian invasions of 1, 95–8, 99–104, 180–88
    see Donbas, Orange Revolution, Revolution of Dignity
Umarov, Mamikhan 157

United Arab Emirates 108, 163
United Kingdom 7, 63, 93, 105, 107, 171
    and Brexit 9, 156, 157, 178
    assistance to Ukraine 183–4
United Nations
    Human Rights Council 9
    Security Council 63, 107, 109, 111, 164, 165
United Russia 89
Utkin, Dmitry 160–61

Van Damme, Jean-Claude 55
Vietnam 13

Wagner Group 115, 124, 154, 159–69, 180, 181, 187
Walesa, Lech 19
war crimes, *see* atrocities
"War on Terror" 9, 50, 67, 108, 150, 157
Warsaw Pact 13, 19, 20, 21, 64

Yanderbiev, Zelimkhan 157
Yanukovych, Viktor 75, 81, 86, 91–5, 96, 97, 153
Yazidis 113
Yeltsin, Boris 2, 8, 25–9, 30, 34–5, 41, 43, 44, 49, 52, 63, 64, 83
    Chechnya 31–3
    Dagomys accord 59–61
    Elections 17, 39
    Nagorno-Karabakh 132, 133, 135, 138
    reforms 12
Yemen 107, 164
Yerevan 22, 135, 140, 142
Yushchenko, Viktor 74, 85, 86, 90, 91
Yugoslavia 46, 158
Yukos 33

Zachistka 31, 48–50
Zakharchenko, Alexander 102
Zelensky, Volodomyr 99, 121, 169, 176–8, 180–81, 189
Zhirinovsky, Vladimir 32, 72
Zuma, Jacob 114
Zyuganov, Gennady 4, 32, 47